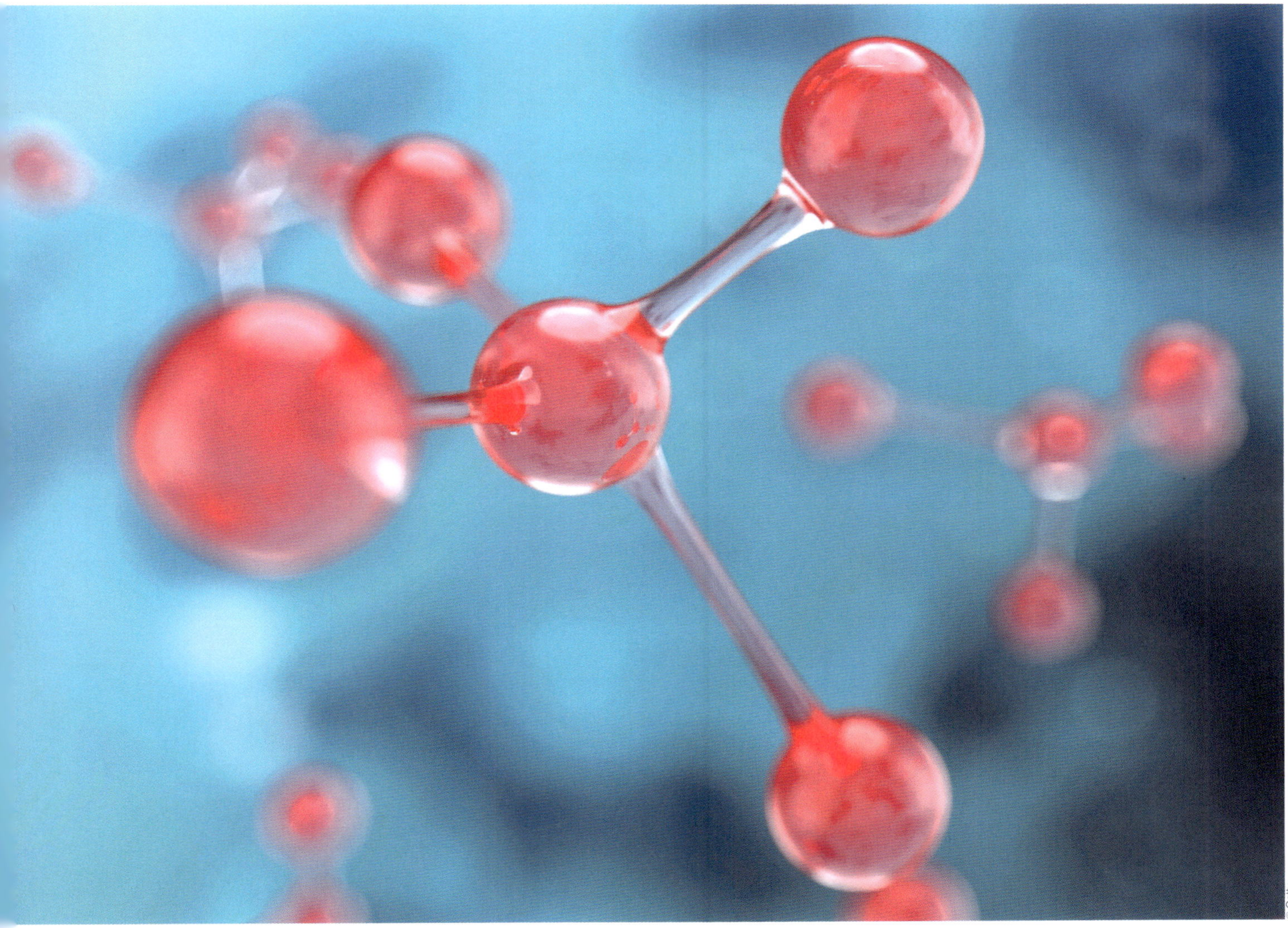
© Getty

Welcome

Chemistry is at the crux of all matter. Everything in our universe is made up of atoms. The way in which the components of these microscopic building blocks are structured defines the substances they make up, their properties and behaviours and the way in which these elemental chemicals interact with one another to create new forms of matter. Inside the pages of *Understanding Chemistry*, you will discover how chemical processes shape our modern lives – from fuelling car engines and powering smartphones to creating our favourite alcoholic beverages and washing our hair.

Contents

60

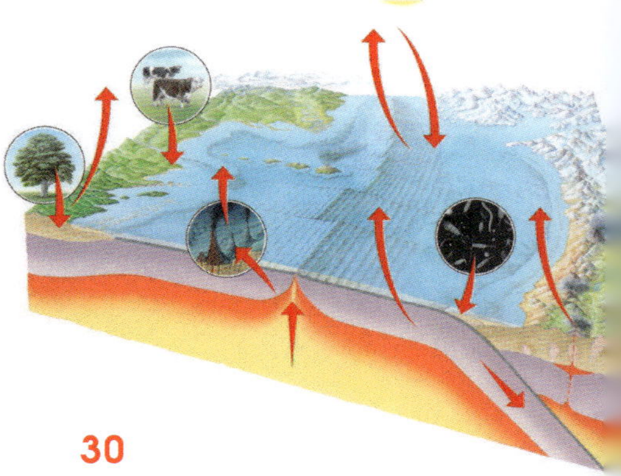
30

10

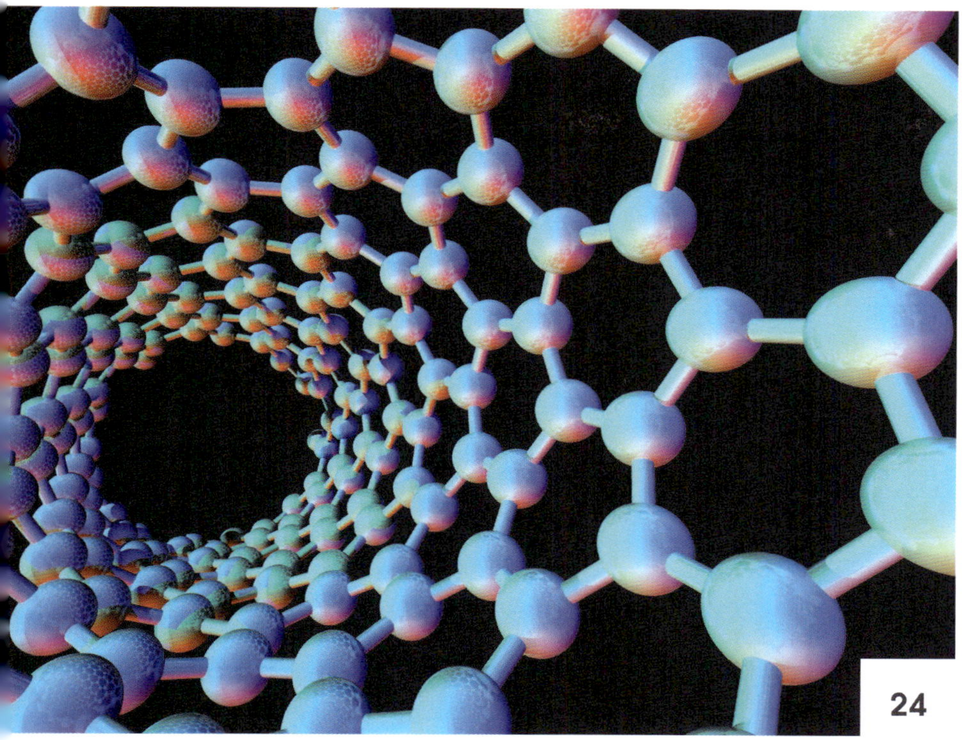

24

74

14

112

50

The power of atoms

Atoms are the ultimate construction kit. With a big enough collection, you can build everything from Venus de Milo to Venus the planet...

Everything in the universe is made of countless tiny atoms that are joined into different structures, essentially like toy building blocks. But instead of snapping together through friction, like plastic blocks, atoms snap together through electrical charge.

At the centre of every atom you'll find the nucleus – a blob of positively charged particles called protons and neutrally charged particles called neutrons. The tiny nucleus is surrounded by even smaller negatively charged particles called electrons. Usually, there is an equal number of protons and electrons. Because of their opposite charge, these protons and electrons are attracted to each other, which is what holds the atom together. With this balance, the atom is electrically neutral.

But these electrons are a fickle bunch: they're not only attracted to their own atom's nucleus – they're sometimes attracted to the nuclei of other atoms as well. In the right situations and conditions, this cross-atom attraction can provide a sort of 'electron glue' that bonds multiple atoms together.

An atom's bonding prospects depend on its proton and electron count and arrangement, which is unique for every element on the periodic table. Electrons surround the nucleus at specific energy levels, called shells. The shell closest to the nucleus is the lowest energy level, and the shell farthest from the nucleus is the highest energy level. Each shell can hold a limited number of electrons. For example, the lowest-level shell holds a maximum of two electrons, and the next level holds up to eight electrons. To achieve maximum stability, electrons move to the lowest possible energy level that has available openings.

The critical factor in chemical bonding is t number of openings in an atom's outermost shell, called the valence shell. When there is right combination of openings, electrons can jump from one atom to another, two atoms share an electron, or many atoms can share a cloud of electrons. Atoms are more stable when their valence shells are full, so electro will readily move in ways that form complet valence shells.

When multiple atoms bond together, they form molecules. Molecules can consist of ma identical atoms – atoms of the same element or they can include atoms of multiple eleme A multi-element molecule is called a compo Collectively, vast numbers of molecules form the wide variety of materials around us. The

full power, trillions of protons will race around the C accelerator ring 11,245 times a second

ucture of these individual molecules, ng with the way that those molecules together, ultimately dictates how any terial both feels and behaves.

Broadly speaking, there are three styles organisation: gases, liquids and solids. gases, molecules move about freely. liquids, molecules fit together loosely, ling over one another like marbles in a wl. In solids, meanwhile, molecules are anged in more rigid structures, and so n't move as freely.

Within these groups, different nbinations and arrangements of atoms ult in an incredible range of qualities l behaviours. Even limited to a set of ntical atoms, structural changes can ke huge differences. For example, npare diamonds and graphite. Both are angements of carbon atoms, but you n't see anyone proposing on bended e with a pencil.

n a diamond, strong covalent bonds atoms in a rigid lattice framework. result is one of the hardest, toughest terials in the world. In graphite, on the er hand, carbon atoms are arranged in yered structure, with very weak bonds ween layers – so weak that touching encil to paper is enough to break m. The ability to combine and arrange ms into different structures yields ually unlimited possibilities. Scientists engineers have already developed ntless thousands of novel materials, we're nowhere close to exhausting the ential combinations.

he chemical reactions involved in ombining atoms can prove useful themselves. For example, fire is the result of a chemical reaction between the chemical compounds in wood (or some other fuel) and oxygen in the atmosphere, triggered by intense heat. Burning wood produces char and gaseous compounds of hydrogen, carbon and oxygen. As the gases start to heat up, the compounds will break apart, and the atoms recombine with oxygen in the air to produce water, carbon dioxide, carbon monoxide and nitrogen; this releases a great deal of energy in the form of heat and light in the process.

Between forming new materials and producing usable energy, the manipulation of atoms has always been at the heart of human technology – even at the stage when we had no clue that atoms existed. In recent years, scientists have managed to make new atoms, forming 20 elements not observed in nature by combining existing nuclei into new super-heavy nuclei. At the moment these manmade atoms quickly fall apart, but stable variations may not be far off. In the 20th century, humans unlocked the internal energy of atomic nuclei for the first time, yielding both nuclear power plants and bombs.

Today, physicists are investigating the even smaller components – quarks, leptons and bosons – that make up atoms. At this still mysterious level, new findings could fundamentally redefine our understanding of the universe.

Anatomy of an atom

What are the fundamental parts that make an atom?

Shells

Electrons can only exist in set energy levels, commonly called shells. Each shell has openings for a limited number of electrons.

Electrons

Electrons are very small, negatively charged particles that move quickly around the atom's nucleus.

The nucleus

The centre of the atom, and almost all of its mass, is the nucleus. The nucleus is made up of protons and neutrons.

Protons

Protons are positively charged particles in the nucleus. All elements are defined by how many protons they have.

Neutrons

Neutrons – particles with no electrical charge – help give atoms their mass. They are slightly bigger than protons.

The 'Fat Man' nuclear bomb was dropped on Nagasaki, Japan, on 9 August 1945

Atomic models

Atoms don't follow the rules of Newtonian physics that we see every day in the world around us, making it impossible to visualise what's actually happening at an atomic level. The best that scientists can do is create theoretical models that give us a general conceptual comprehension of what's going on. Here are some of the noteworthy models that greatly advanced our atomic understanding...

Thomson's 'Plum Pudding' model (1904)

English physicist JJ Thomson discovered the electron as far back as 1897, showing for the first time that atoms had smaller constituent components. To account for the atom's overall neutral charge, Thomson theorised – in his 1904 model – that the negatively charged electrons must sit in a regular pattern within a uniformly distributed positive charge, like raisins in a plum pudding.

Rutherford's nuclear model (1911)

New Zealand-born physicist Ernest Rutherford, who studied at Cambridge University, disproved Thomson's model, when he demonstrated the existence of a positively charged atomic nucleus. Rutherford proposed the atom was like a miniature solar system, with a relatively massive, Sun-like nucleus at the centre, orbited by much smaller planet-like electrons.

Bohr's shell model (1913)

In classical mechanics, any charged particle moving in a curved path emits radiation. Consequently, in Rutherford's model, electrons would lose energy and collapse into the nucleus. Danish physicist Niels Bohr proposed electrons moved in a different type of orbit. He theorised electrons surrounded a nucleus in fixed energy levels (shells) and only emitted radiation when they 'jumped' from shell to shell.

What are elements and compounds?

Elements are substances made entirely of one type of atom. Each element is defined by how many protons are in a single atom of that element. For example, every hydrogen atom has just one proton, while every gold atom has 79 protons, and so on. In the right circumstances, atoms of different elements can join together to form a chemical compound. The bonds that hold compounds together result from various movements of electrons. Here are two examples:

Ionic bond

1 Ionic bond
Ionic bonds form when an electron jumps from one atom to another, resulting in two electrically charged atoms, called ions.

2 Valence shell
Electrons travel in set energy levels called shells. Each shell has a limited number of openings for electrons. The number of openings in the outermost level, known as the valence shell, determines how an atom can form bonds.

3 Sodium atom
The sodium atom's valence shell has only one electron, leaving seven openings.

4 Electron leap
To achieve overall stability, the spare electron from the sodium atom leaps to fill the chlorine atom's valence shell.

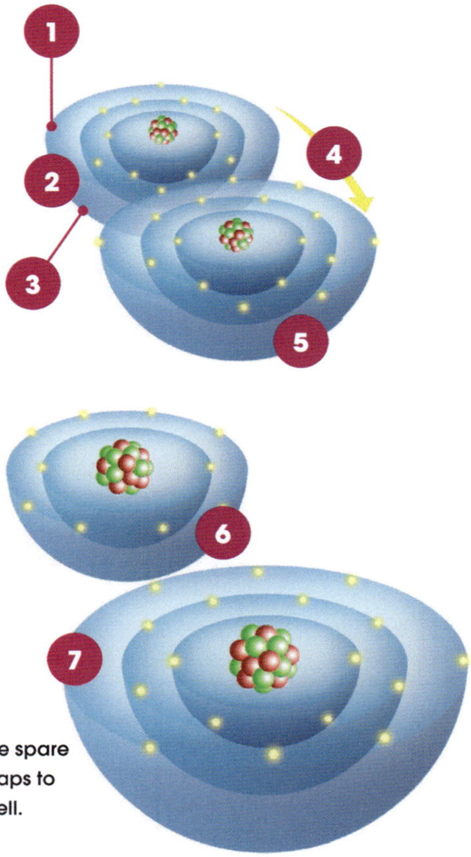

5 Chlorine atom
A chlorine atom's valence shell holds seven electrons, leavin
only one opening.

6 Sodium cation
The sodium atom now has ten electrons and 11 protons, makin
a cation – an atom wit
a net positive charge.

7 Chlorine anion
The chlorine atom now has 18 electrons a
17 protons, making it a
anion – an atom with a
net negative charge. T
opposite charges bon
the two atoms togethe
to form the compoun
sodium chloride, more
commonly known as table salt.

Covalent bond

1 Covalent bond
Atoms can also form compounds by sharing electrons between them, in covalent bonds.

2 Electron pairs
Each of the three hydrogen atoms shares its original single electron and one of the original nitrogen electrons, collectively forming the compound ammonia.

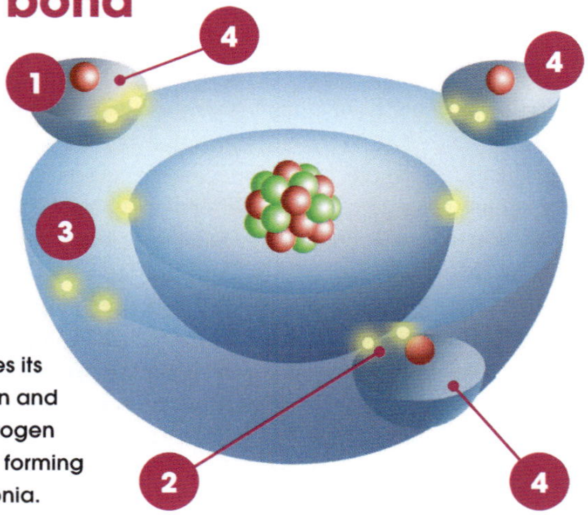

3 Nitrogen atom
A nitrogen atom has five electrons in its valence shell, leaving three openings.

4 Hydrogen atoms
Each of these three hydrogen atoms has a single electron in its valence shell, leaving one opening.

Bombarding atoms together leads to the dislodging of subatomic particles

rom stars to space dust, everything
n the universe is made up of atoms

Radioactive decay explained

Most atoms are highly stable, meaning the nucleus will always hold together, barring extreme circumstances. ut in some atoms, the energy that binds the nucleus vill eventually fail in a process called radioactive decay – the spontaneous disintegration of the nucleus.

The most notorious unstable atoms are elements with very high number of protons, such as uranium (92 protons). But some lighter elements, eg carbon, can be nstable and radioactive when they have too many or oo few neutrons. Neutron-count variations are called sotopes. For example, while garden-variety carbon-12 six protons and six neutrons) is stable, carbon-14 (six protons and eight neutrons) is radioactive.

Radioactive decay results in the ejection of ubatomic particles from the nucleus. In alpha adiation, the atom ejects two protons and two eutrons. In beta radiation, a neutron turns into a roton, ejecting a neutrino particle and a free electron, alled a beta wave. In gamma radiation, the nucleus eleases energy in the form of a photon. Energy from he ejected particles can damage and mutate human NA, sometimes causing cellular changes, ie cancer.

The nucleus material that's left forms a 'daughter tom'. When the proton count has changed, the aughter atom will be different from the original atom. arbon-14 decays into nitrogen, for example.

How to split an atom

Add this to the pile of mind-bending atom qualities: an atomic nucleus has less mass as a whole than its protons and neutrons would have separately. How is this possible? Well, when the nucleus is formed, some of the mass of its constituent parts changes into energy that binds the protons and neutrons together. In other words, there's high potential energy locked up in the nucleus of the atom.

It's possible to release this energy, and actually harness it, by splitting specific types of atoms apart into multiple fragments – a process known as nuclear fission. All you need to break apart a uranium-235 atom is a slow-moving free neutron. The uranium atom will absorb the free neutron, the extra energy makes the uranium nucleus highly unstable, and the atom splits into two smaller atoms and two or three free neutrons. The potential energy in the nucleus is released as kinetic energy, in the form of these particles moving at great speed. The resulting free neutrons, in turn, can break apart other uranium-235 atoms, leading to a chain reaction.

A power plant controls the reaction and harnesses the heat of this kinetic energy in order to generate steam that turns turbines. In contrast, in an atomic bomb the reaction is allowed to go unchecked, in order to generate a massive explosion.

You can also tap into this energy through nuclear fusion – the combining of two nuclei into a new nucleus. Nuclear fusion generates the energy of stars and hydrogen bombs. However, nobody has been able to harness it effectively as a power source yet.

> "All you need to break apart a uranium-235 atom is a slow-moving free neutron"

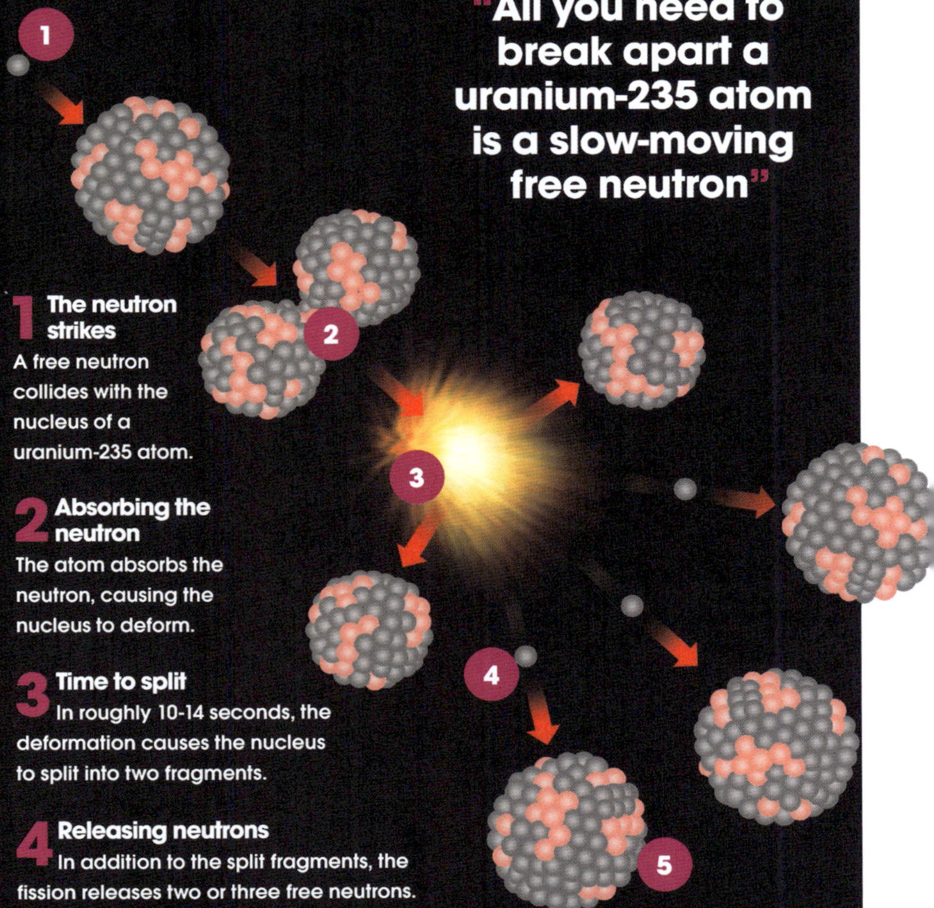

1 The neutron strikes
A free neutron collides with the nucleus of a uranium-235 atom.

2 Absorbing the neutron
The atom absorbs the neutron, causing the nucleus to deform.

3 Time to split
In roughly 10-14 seconds, the deformation causes the nucleus to split into two fragments.

4 Releasing neutrons
In addition to the split fragments, the fission releases two or three free neutrons.

5 Neutrons strike back
The free neutrons from the reaction encounter other uranium-235 nuclei, beginning the process all over again, leading to a chain reaction.

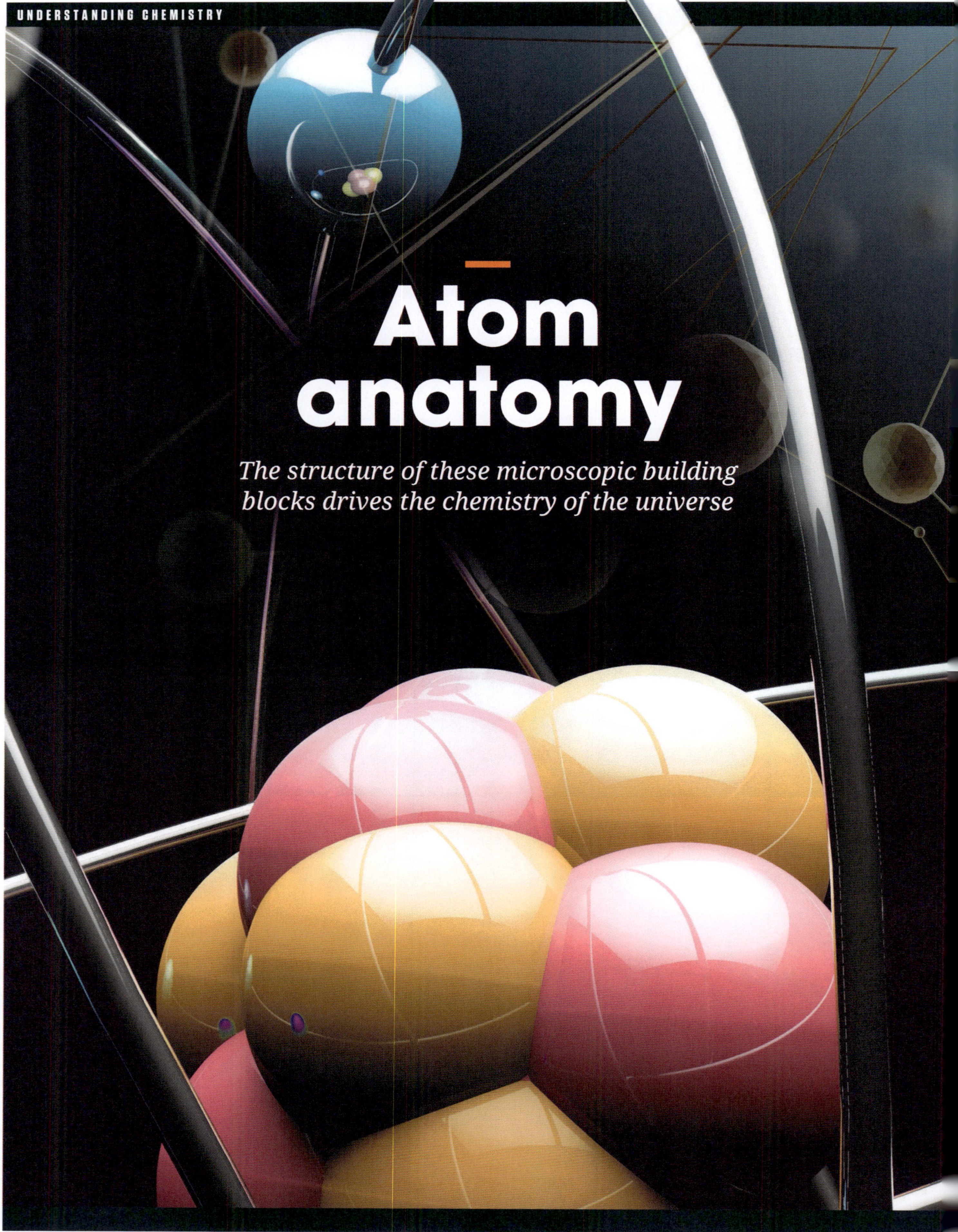

Atom anatomy

The structure of these microscopic building blocks drives the chemistry of the universe

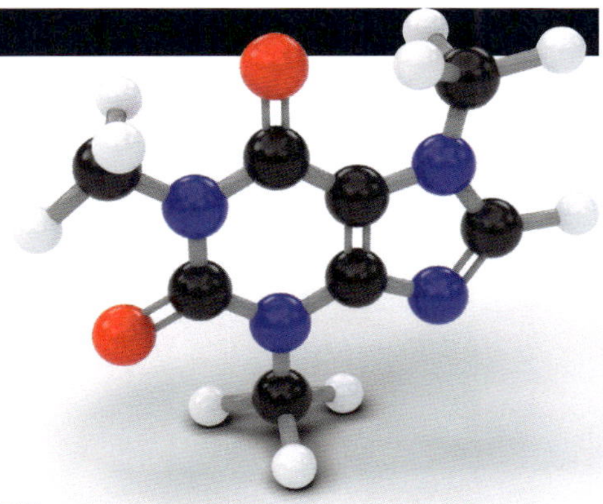

Different elements can combine together to make complicated compounds

A toms are the building blocks of everything around you, from the cells in your body to the cup of tea in your hand. Made from elementary particles and ld together by fundamental forces, they drive e chemistry of the universe.

Each one has a tiny, positively charged cleus, balanced by a cloud of negatively arged electrons. These determine the atom's operties. Split an atomic nucleus open and u'll find subatomic particles called protons and utrons. The simplest atom, hydrogen, has just e proton and no neutrons. But as you move up ough the periodic table the number of protons es one by one. With each extra proton, the naviour of the atom changes.

Protons have a positive charge, so for every extra one added, the atom will also gain another negative electron. These electrons cluster in fixed zones around the atomic nucleus, known as orbitals or shells. Each shell can only hold a certain number of electrons. This means that when a shell fills up, the next electrons then have to start a new one.

Atoms are most stable when their electron shells are full, but most elements have electrons to spare. This is the driving force behind the

"Split an atomic nucleus open and you'll find subatomic particles"

se are sugar crystals magnified 1,250 times by a scanning electron microscope

What your tea's made of

There are more atoms in a cup of tea than cups of water in all the oceans

Mixture

Your cup of tea is a mixture; a combination of different molecules and compounds not linked by chemical bonds. It contains water, as well as the molecules from the tea itself: polyphenols, amino acids, enzymes, pigments, volatiles and caffeine. Perhaps milk is there too, which contains fats, proteins and sugars. And maybe also table sugar.

Scale: 10 centimetres

Compound

If you stuck a powerful microscope into your tea, you'd start to see the compounds that make up the mixture. Compounds are substances made from two or more different elements bonded together. Caffeine has the chemical formula C8H10N4O2. It contains eight carbon atoms, ten hydrogen, four nitrogen and two oxygen.

Scale: 0.78 nanometres

Molecule

Break caffeine into pieces and you'd end up with molecules. These contain just one type of atom and are the smallest units that can take part in chemical reactions. If you broke the caffeine compound up, its oxygen atoms would pair together, forming an oxygen molecule with two identical atoms.

Scale: 292 picometres

Element

Zoom in on an oxygen molecule and you'd see the individual oxygen element. Elements are substances that contain only one type of atom. There are over 90 naturally occurring elements on Earth, and even more made synthetically in laboratories. Oxygen is element number 8.

Scale: 60 picometres

Atom

One final zoom would reveal the subatomic particles inside the atoms: positive protons, neutral neutrons and negative electrons. The numbers of each give different elements their distinctive chemical properties. An oxygen atom contains eight protons, eight neutrons and eight electrons.

Scale: 1 femtometre

The birth of atoms

he Big Bang created the universe roughly 3.8 billion years ago. In the first moments it vas so hot and dense that no particles could urvive. Whenever quarks and electrons burst nto existence, they immediately broke apart. But as the universe expanded and tarted to cool, these elementary particles became more stable. Then quarks started to come together to form protons and neutrons. Within minutes, these were combining to orm the first atomic nuclei.

It took another 380,000 years for the niverse to cool enough to allow complete ydrogen atoms to form, and then the real magic began to happen. Gravity started to ull the atoms into clumps, and as those lumps grew the pressure inside started to uild. The atomic nuclei got so close together that they started to fuse, and the first stars burst into life. These vast nuclear furnaces have been fusing atomic nuclei ever since, creating all the different elements we see around us today.

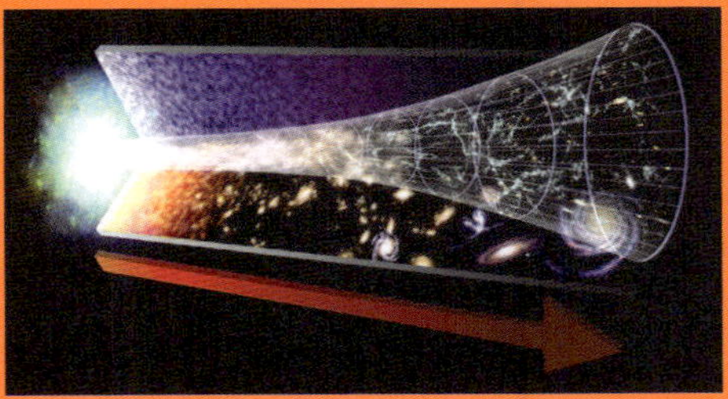

Hydrogen atoms were the first to form after the birth of the universe

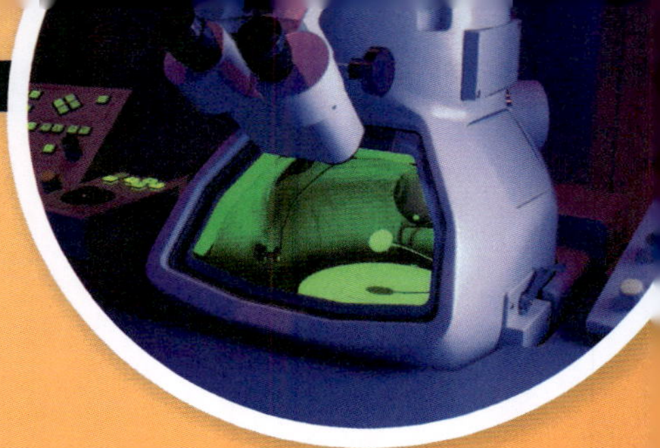

The highest-resoluti
electron microscopes c
zoom in on single ato

Inside an atom

Lifting the lid on the subatomic particles that make up the atoms in the air we breathe

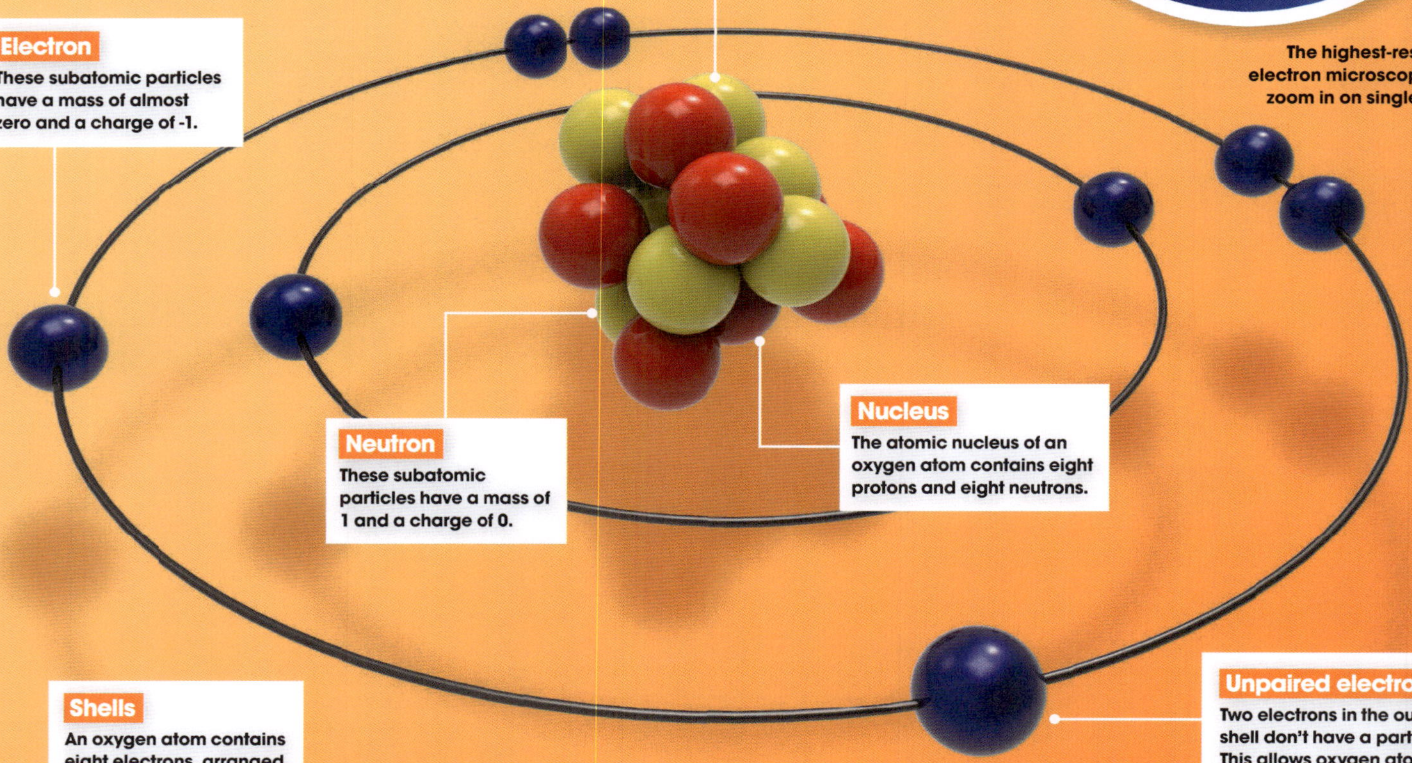

Proton
These subatomic particles have a mass of 1 and a charge of +1.

Electron
These subatomic particles have a mass of almost zero and a charge of -1.

Neutron
These subatomic particles have a mass of 1 and a charge of 0.

Nucleus
The atomic nucleus of an oxygen atom contains eight protons and eight neutrons.

Shells
An oxygen atom contains eight electrons, arranged into two orbital shells.

Unpaired electrons
Two electrons in the outer shell don't have a partner. This allows oxygen atoms t form chemical bonds.

ATOM FACTS

118
Number of elements in the periodic table

90
Number of elements that exist in nature in appreciable quantities on Earth

50 million
Chemicals have been discovered or made by humans

75%
Most of the matter in the universe is hydrogen

2.6 secs

A new chemical is being made with speedy regularit

The number of chemicals currently in use in industry
7,700

tars made all of the elements
eavier than hydrogen and helium

emistry we see around us. To complete half-
ed shells, atoms form chemical bonds. They
n either share their electrons to form covalent
nds, or they can steal electrons from one
other to form ionic bonds.

Covalent bonds are common inside our own
dies. Carbon atoms need four electrons to
nplete their outer shell, so they share with
to four other atoms, including other carbons.
is makes them perfect for forming large
logical molecules like DNA.

onic bonds are more commonly found in the
und. They happen when a non-metal steals
ctrons from a metal, becoming negatively
rged. At the same time, the metal becomes
sitively charged. These ions attract one another, sticking together in repeating structures, like salt crystals.

> "They can share their electrons to form covalent bonds, or they can steal electrons"

Bonds between atoms can break and reform, allowing the elements to perform chemical tricks. They combine to produce larger compounds, decompose to form smaller ones, or swap one element for another to produce a chemical with different properties. These dynamic interactions are happening all around us, all the time.

Smaller still

The protons and neutrons that make up the atomic nucleus aren't the smallest components of matter. These are the 'elementary particles' – quarks and leptons. There are six types of quark, arranged into pairs: up and down, charm and strange, top and bottom. They have a partial electrical charge and a 'colour charge' of red, green or blue. There are also six types of lepton: electron, muon, tau and their associated neutrinos. It takes four particles to make up the simplest hydrogen atom: two up quarks and one down quark for the proton and one electron to orbit around the outside. The most common form of hydrogen has no neutrons.

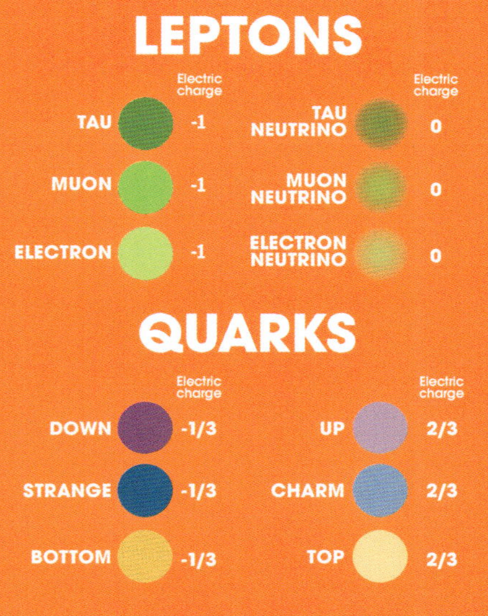

There are six quarks and six leptons, each with different properties

tomic evolution

Our understanding of atomic structure has come a long way in the past 200 years

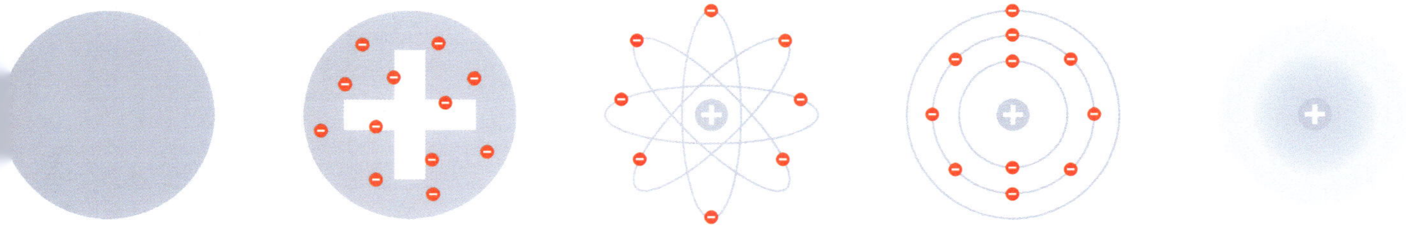

03 Solid sphere

n Dalton was the first to
cribe atoms. The word came
n the ancient Greek 'a',
aning 'not', and 'temnein',
aning 'to cut'. He thought
they were indivisible.

CURACY

on recognised that different
s of atom had different
erties. But he didn't realise
contained smaller
atomic particles.

1904 Plum pudding

J.J. Thomson discovered electrons and made a new model of the atom to include them. He stuck them inside a sphere of positive charge, like fruits in a cake.

ACCURACY

This model recognised that atoms contained smaller components, including electrons. But the structure was incorrect; the atomic nucleus was missing.

1911 Nuclear

When Ernest Rutherford tried firing tiny particles at atoms, most passed through. This revealed that most of the atom is empty space, with a small nucleus at the centre.

ACCURACY

A positive nucleus surrounded by negative electrons is close to our modern understanding of atoms. But Rutherford hadn't quite captured how electrons move.

1913 Planetary

Niels Bohr's planetary model describes electrons travelling in orbits called 'shells'. They are a fixed distance from the nucleus and have a fixed energy level.

ACCURACY

The planetary model explains why small elements absorb and emit certain frequencies of radiation. However, it runs into problems as elements start to get heavier.

1926 Quantum

Rather than think of orbitals as circular paths, Erwin Schrödinger imagined them as probability clouds. There is a 90 per cent chance of finding an electron somewhere inside its orbital.

ACCURACY

This model is the most up to date. Thinking of electrons as waves in clouds of probability helps to explain the behaviour of subatomic particles.

Your guide to the elements

How these chemical building blocks make up life, the universe and everything

All of the 118 elements in the periodic table are made from the same three key ingredients – protons, neutrons and electrons. The protons and neutrons make up the nucleus at the centre of each atom, while the electrons whizz around the outside and make chemical bonds with other atoms. The identity of each atom is determined by the number of protons in its nucleus, known as the atomic number. Hydrogen has one, helium has two, lithium three, and so on. The periodic table lists the elements in this order.

Protons are positively charged, while electrons are negatively charged; an atom will have an equal number of each. The electrons are arranged in 'shells' around the nucleus. Each row of the periodic table represents a new layer of electron shells, and each column represents how full the outer shell is. For example, elements in the first column of the periodic table – including lithium and sodium – have just one electron in their outer shell, while those in the second column – such as beryllium and magnesium – have two.

The number of electrons in the outer shell affects how the element behaves, so those in the same column have similar properties. Atoms like to have a full outer shell of electrons, so those with one or two extras are keen to give them away, and those with gaps want to fill them up. If you drop any of the elements from the first column into water, they will fizz, flame or even explode as they race to share their spare electron with other atoms, but if you did the same with the elements in the last column, nothing

How to read the periodic table

There are three key pieces of information about each element to look out for

Atomic number

The atomic number tells you how many protons each atom has. This is different for every element.

Chemical symbol

Each element has a one- or two-letter symbol. It is often based on Latin, and may not relate to the English name.

Atomic mass

Some tables also give the atomic mass, which corresponds to the total number of protons and neutrons in the atom.

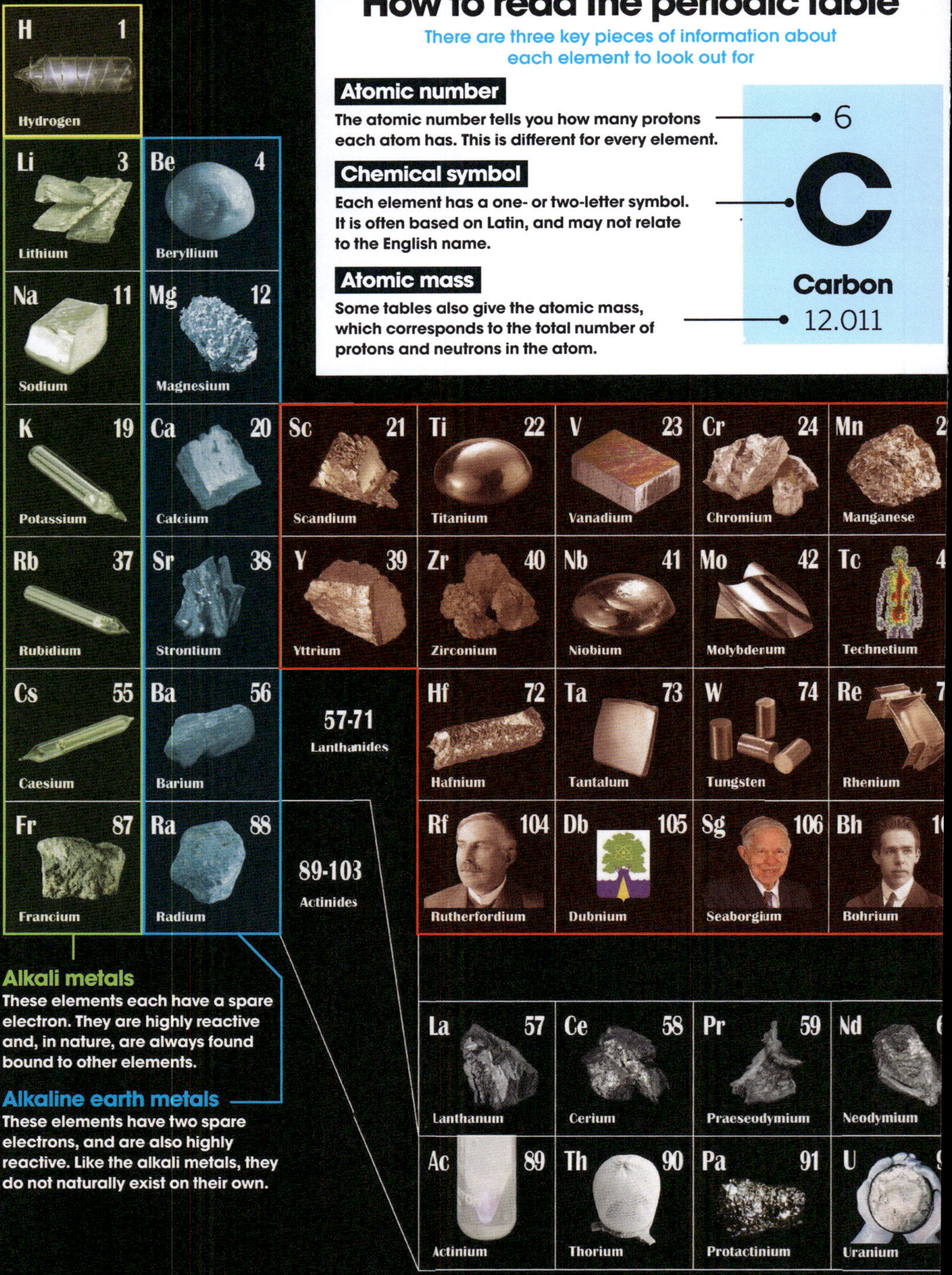

Alkali metals
These elements each have a spare electron. They are highly reactive and, in nature, are always found bound to other elements.

Alkaline earth metals
These elements have two spare electrons, and are also highly reactive. Like the alkali metals, they do not naturally exist on their own.

ould happen. These elements have a ıll outer shell, so don't need to share ıeir electrons with other atoms.

Most of the elements in the periodic ıble occur naturally on Earth, but any lement heavier than lead (number 82) ; unstable and gradually undergoes adioactive decay. Elements heavier ıan uranium (number 92) have to be ıade artificially. Join us as we explore ıe periodic table, and delve into the lements that shape our everyday lives.

Non-metals
The elements in the top-right corner of the periodic table are the non-metals. Most are gases or solids when at room temperature.

Halogens
The halogens are missing one electron from their outer shell, and will react violently with the alkali metals to form salts.

Noble gases
These elements have a complete outer shell of electrons and do not react with other elements.

Transition metals
The elements in this block have more than one partially filled electron shell, giving them interesting chemical properties.

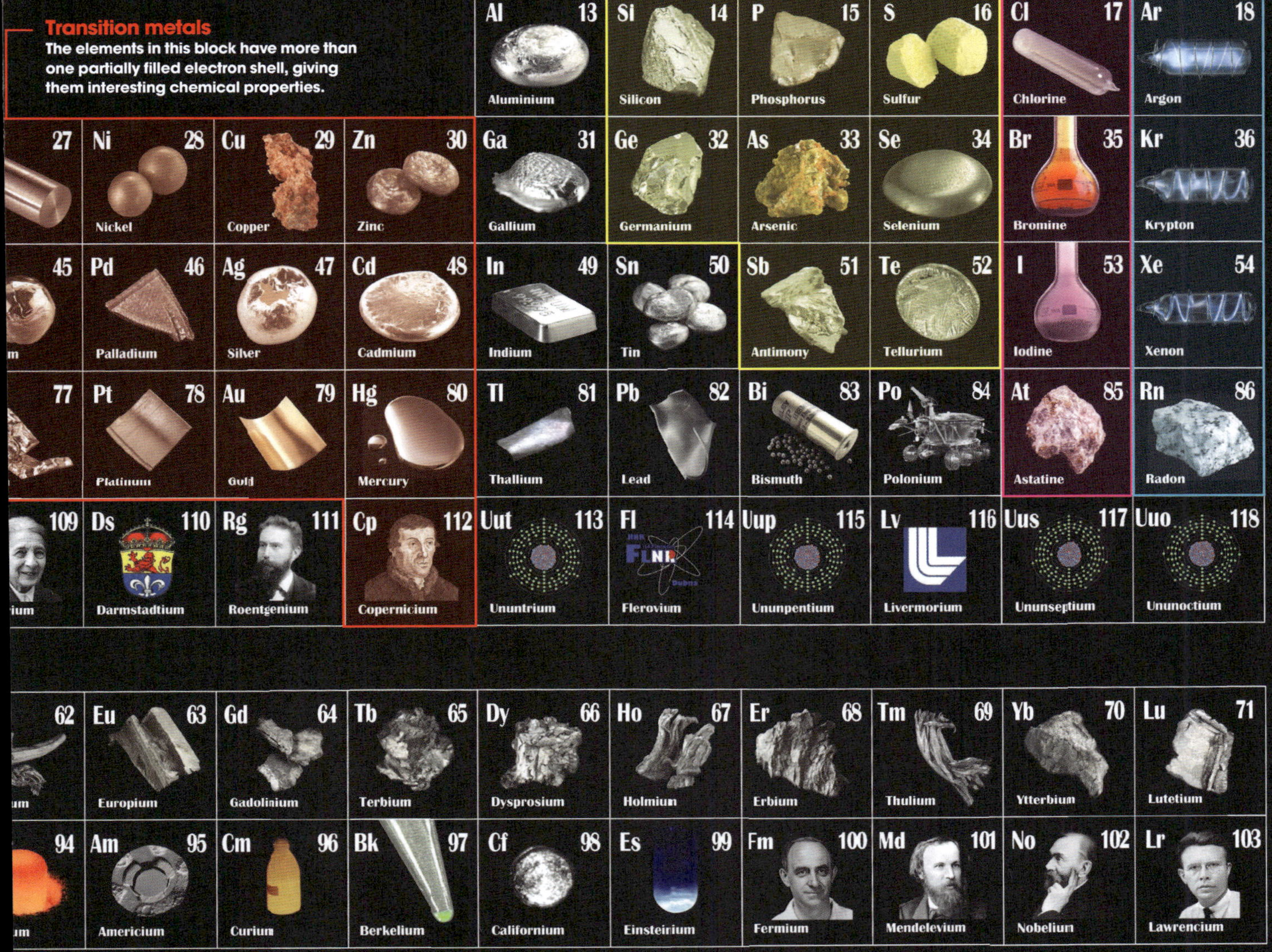

Everyday elements

Look around and you'll discover dozens of different elements

Gas-discharge lamps

These lamps typically contain neon, argon, krypton, or xenon, which are noble gases. When an electric current is passed through the gases, they become excited, and when they drop back down to a normal energy level, they release photons of visible light.

Ne | Ar | Kr | Xe

Coins

In the UK, 1p and 2p coins are made from copper-plated steel (iron and carbon), 5p and 10p coins are nickel-plated steel, 20p and 50p coins are cupro-nickel (copper and nickel), and £1 and £2 coins are nickel-brass (copper, nickel and zinc). These elements are cheaper than gold or silver, and durable too.

Fe | C
Cu | Ni | Zn

Heat resistant glas

Borosilicate glass, which is found in
kitchen and in the lab, contains at least fi
per cent boron oxide. Boron has a hi
melting point, which helps the glass to re
thermal shock, going from hot to cold a
back again without shatteri

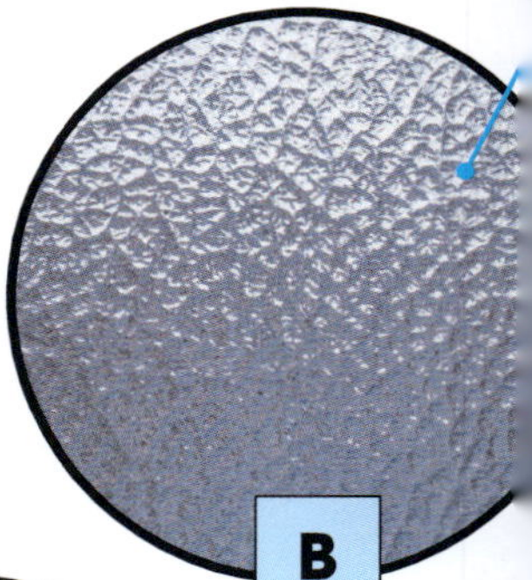

B

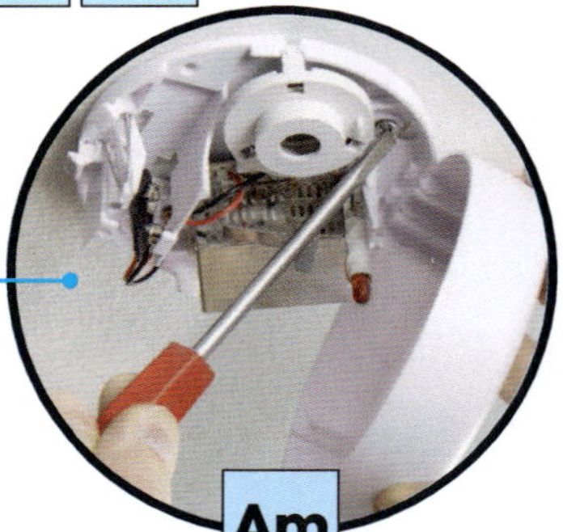

Am

Smoke detectors

Many smoke detectors contain small amounts of americium. This radioactive element releases alpha particles, which 'knock' electrons away from gases in the air and towards a positively charged plate in the smoke detector, generating a current. When smoke gets in the way, the current stops and the alarm sounds.

Sr | Li | C
Na | Ba | C

Firewor

The colours of fireworks are produ
using various combinations of eleme
which burn with different coloured flan
Strontium and lithium salts burn red
calcium salts burn orange, while sod
salts burn yellow, barium salts green,
copper salts blue. Purple can be made
mixing strontium and cop

Phone ingredients

The shiny metal, glass and plastic exteriors of smartphones can contain lots of different elements, including aluminium, magnesium, carbon, and oxygen, but this is just the tip of the iceberg; the circuitry, battery, cameras and speakers contain dozens more. Silicon chips coated with antimony and arsenic sit beside batteries containing lithium and cobalt, and the features that we take for granted, like colour screens, are made possible by rare earth metals like terbium and europium. Finding uses for all of these elements is a real achievement, but as demand for electronics soars, our limited resources could start to run out.

In | Sn | O

Indium tin oxide film

Three elements – indium, tin and oxygen – make up the conductive film found inside touch screens.

Al | Si | O | K

Aluminosilicate glass

This tough glass is made from a mixture of aluminium, silicon, oxygen and potassium.

Li | Co | O

Lithium ion battery

The positive electrode contains lithium, cobalt and oxygen, and the negative electrode is made from carbon.

La | Gd

Rare earth circui

Smartphone circu
contains rare earth me
like lanthanum, gadolin
and dysprosi

Cu | Ag | Au

Precious me wiri

Copper, silver, g
and tanta
are use
micro-electr
compon
and wi

Elements, mixtures and compounds

What are the differences between these configurations of atoms?

All matter is made up of atoms – these are tiny particles that cannot be seen through the lense of a conventional microscope.

Atoms make up the table elements, and all elements are made up of the me parts: protons, neutrons and electrons. The oms of a particular element are all the same as ch other, as they all contain the same number protons in their nuclei.

A compound contains atoms of two or more ferent elements, which are chemically joined gether. Water is a compound that contains two drogen atoms and one oxygen atom.

A mixture is a substance consisting of ferent atoms, molecules or compounds that en't chemically joined. Air, for example, is a xture of nitrogen, oxygen, argon and others.

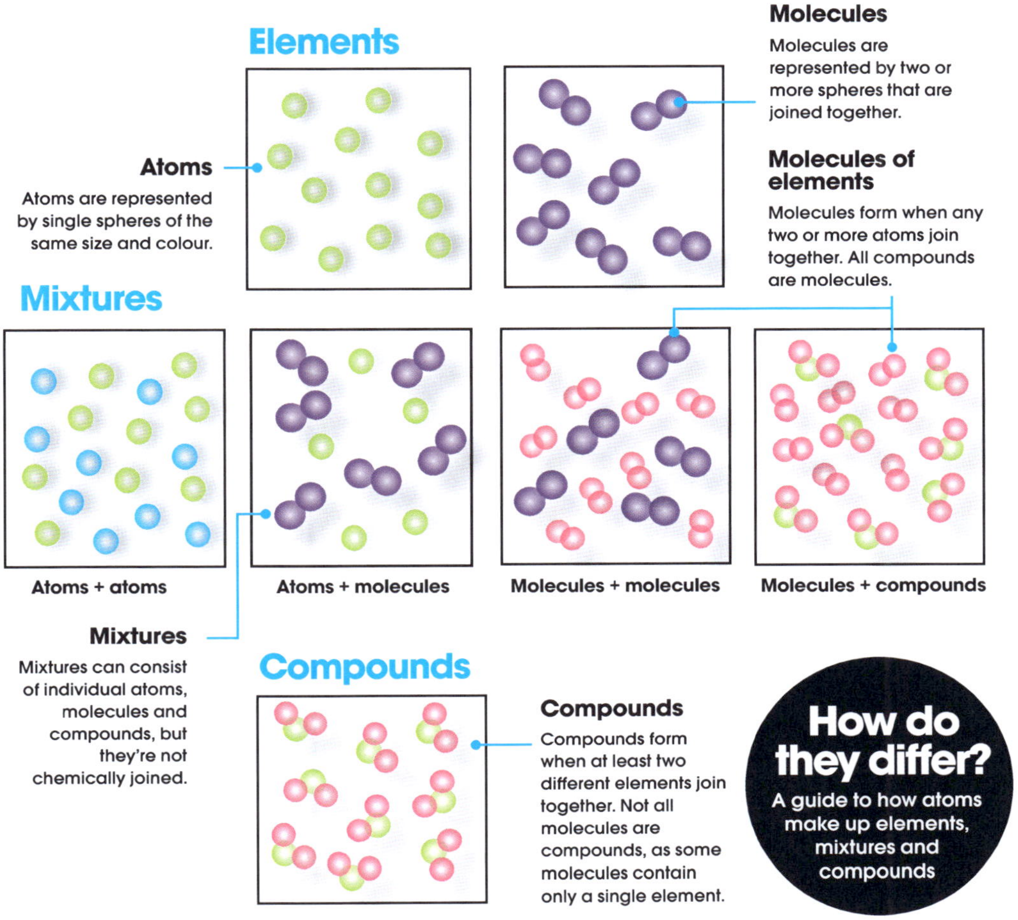

ubstances in mixtures are not chemically bound together"

Atomic energy levels

low atoms absorb or release energy depending on their electron arrangement

n atom's electrons orbit he nucleus at specific nergy levels. The easiest vay to imagine this concept is with a simplified tomic model, in which lectrons orbit the nucleus ke planets orbit a star.

Each defined orbit is nown as an energy level. lectrons must have a ertain amount of energy to rbit within each level, and his amount increases the urther away an energy level from the nucleus.

Electrons can gain or lose energy. If an electron absorbs enough energy in order to move up to the next energy level, it is said to be in an excited state. The electrons in a substance can become excited if the matter is heated or if it absorbs pulses of electromagnetic radiation, such as light photons.

Being excited is an unstable state, so the electron eventually returns to its original, lower-energy level. This transition leads to the emission of a photon with the same amount of energy as the difference between the two energy levels it moves between.

The atoms of different elements have specific energy levels, so they will absorb and emit photons of different energies. Measuring the properties of photons absorbed or emitted by a substance can be used to identify what elements it contains.

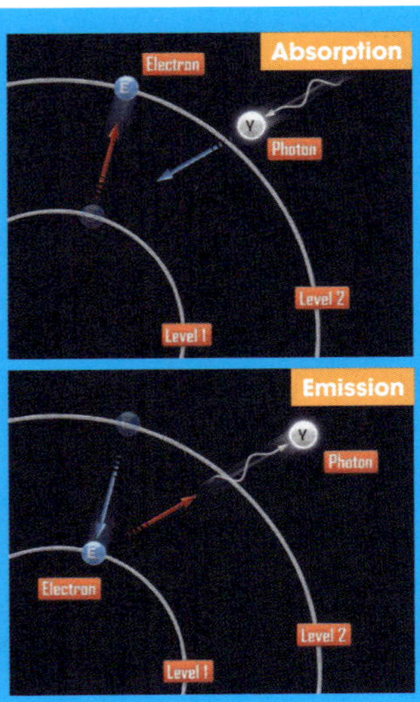

Absorbing a photon of a specific energy enables an electron to move to a higher energy level (top image). When it eventually moves back to its original energy level, a photon will be emitted (bottom image).

You are made of stardust

The elements that make up our bodies were forged inside ancient stars

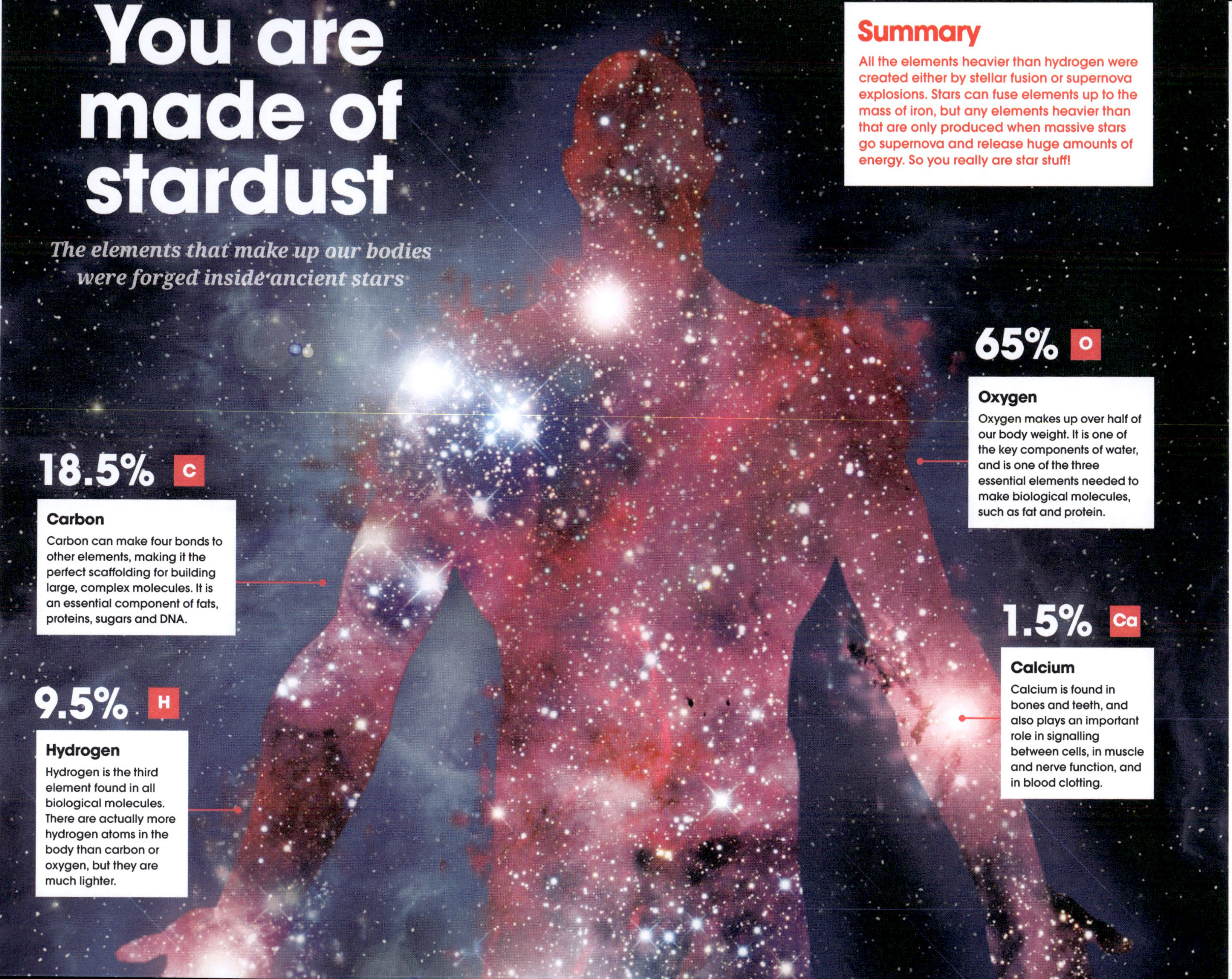

Summary

All the elements heavier than hydrogen were created either by stellar fusion or supernova explosions. Stars can fuse elements up to the mass of iron, but any elements heavier than that are only produced when massive stars go supernova and release huge amounts of energy. So you really are star stuff!

18.5% C

Carbon

Carbon can make four bonds to other elements, making it the perfect scaffolding for building large, complex molecules. It is an essential component of fats, proteins, sugars and DNA.

9.5% H

Hydrogen

Hydrogen is the third element found in all biological molecules. There are actually more hydrogen atoms in the body than carbon or oxygen, but they are much lighter.

65% O

Oxygen

Oxygen makes up over half of our body weight. It is one of the key components of water, and is one of the three essential elements needed to make biological molecules, such as fat and protein.

1.5% Ca

Calcium

Calcium is found in bones and teeth, and also plays an important role in signalling between cells, in muscle and nerve function, and in blood clotting.

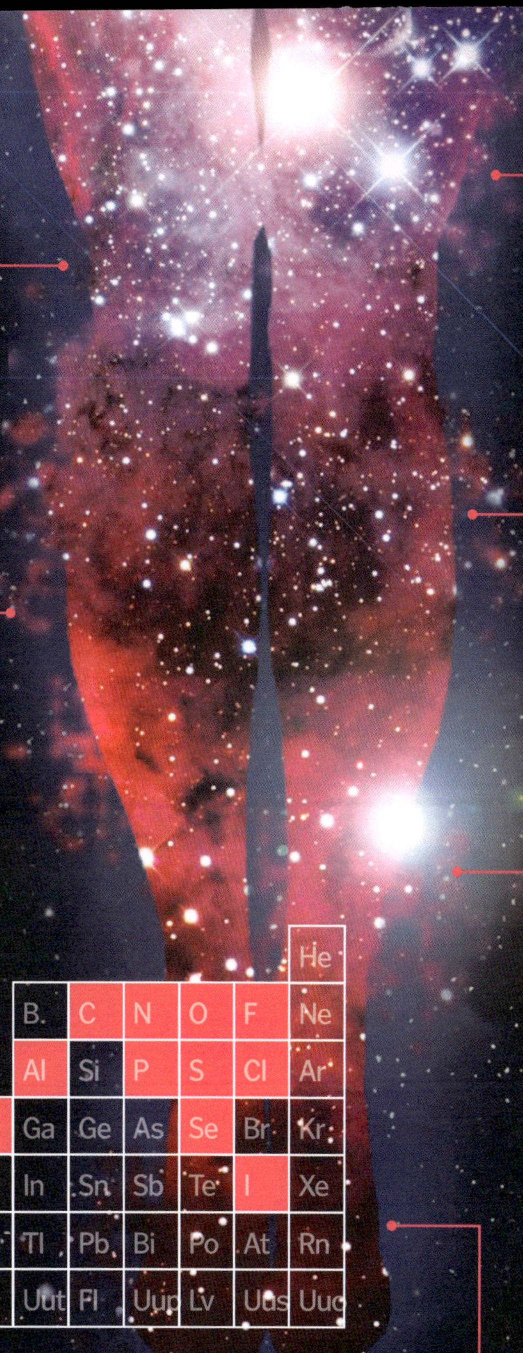

1.1% P

Phosphorus

Phosphorus, like calcium, helps to provide strength to bones and teeth. It is also involved in energy use, and is a vital component in DNA, helping to hold the whole structure together.

0.4% K

Potassium

Potassium ions are found dissolved inside cells and in body fluids. They carry an electric charge, and are used by nerve cells and muscle cells in the transmission of electrical impulses.

3.2% N

Nitrogen

Oxygen, carbon and hydrogen make up the core of all biological molecules, but lots of other elements are used in smaller amounts. Nitrogen is found in both DNA and protein.

0.2% S

Sulphur

Sulphur is found in some of the building blocks of protein. It can make strong bonds to other sulphur atoms, helping to fix proteins into their 3D shapes.

0.2% Na

Sodium

Sodium is another electrolyte that carries charge inside the body. Along with potassium and calcium, it is one of the key elements responsible for normal nerve and muscle function.

The periodic table of the elements

■ Found in the human body – some others are found only in trace amounts

H																	He
Li	Be											B	C	N	O	F	Ne
Na	Mg											Al	Si	P	S	Cl	Ar
K	Ca	Sc	Ti	V	Cr	Mn	Fe	Co	Ni	Cu	Zn	Ga	Ge	As	Se	Br	Kr
Rb	Sr	Y	Zr	Nb	Mo	Tc	Ru	Rh	Pd	Ag	Cd	In	Sn	Sb	Te	I	Xe
Cs	Ba	Lu	Hf	Ta	W	Re	Os	Ir	Pt	Au	Hg	Tl	Pb	Bi	Po	At	Rn
Fr	Ra	Lr	Rf	Db	Sg	Bh	Hs	Mt	Ds	Rg	Cn	Uut	Fl	Uup	Lv	Uus	Uuo

La	Ce	Pr	Nd	Pm	Sm	Eu	Gd	Tb	Dy	Ho	Er	Tm	Tb
Ac	Th	Pa	U	Np	Pu	Am	Cm	Bk	Cf	Es	Fm	Md	No

0.4%

Cl Mg Mn Fe F Co
Cu Zn Se Mo I Li Al

And the rest

There are many other trace elements in the human body, including chlorine, magnesium, manganese, iron, fluorine, cobalt, copper, zinc, selenium, molybdenum, iodine, lithium, and aluminium.

*Percentages are by mass

What is DNA?

Every living creature is a consequence of DNA – discover how it works here

Deoxyribonucleic Acid (DNA) is a chemical library, storing vast quantities of data for billions of years. Passed from one generation to the next, this microscopic structure is the engine of evolution. Every living creature uses DNA as its instruction manual. The genes in DNA tell cells how to make the proteins they need to survive, grow and reproduce. It may be the most famous molecule in the world, but what goes on inside that double helix? At the heart are four chemicals called nucleotides. Each contains a pentagon-shaped sugar molecule, a structure called a phosphate group and a protein fragment called a base.

The sugar and the phosphate group are the same for every nucleotide. They join together side by side to form the backbone of each DNA strand. Like LEGO bricks, they only connect one way round; the phosphate of one nucleotide slots into the sugar of the next. This means that DNA strands have a definite 'up' and 'down'. The top is called the five prime (5') end, and the bottom the three prime (3') end.

The two strands of a DNA helix face in opposite directions, one running five prime to three prime, and the other three prime to five prime. They twist together in a right-handed helix. If you lo closely you'll notice that the alternating twists are uneven: one big (the major groove) and on small (the minor groove).

The bases connect the two DNA strands together, much like the rungs of a ladder. They form pairs, known as base pairs, linked by strong interactions called hydrogen bonds. There are four bases – adenine (A), cytosine (C), guanine (G) and thymine (T) – and they

How it replicates

It takes an hour to turn one DNA double helix into two identical copies

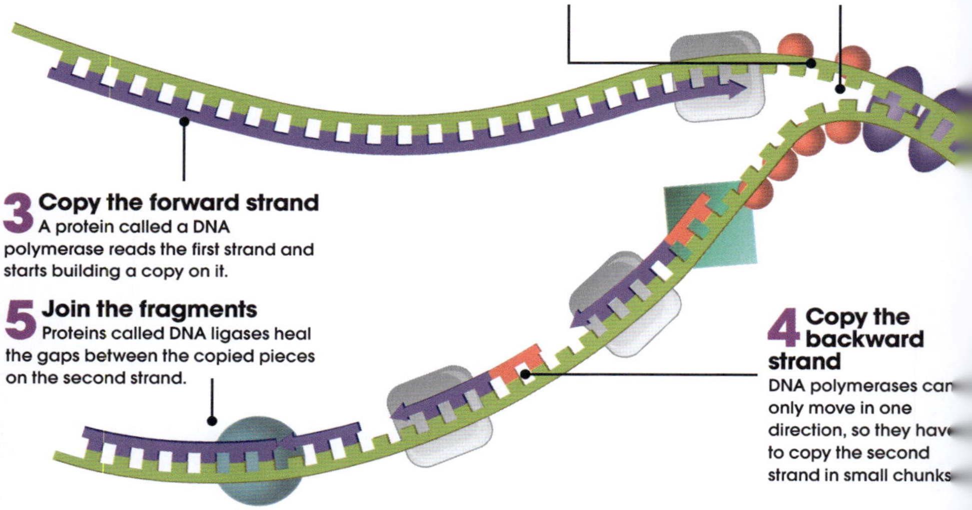

1 Unzip the helix The two strands of the DNA helix unzip to expo the bases inside.

2 Stabilise the strands Helper proteins cling to both strands to stop them from sticking back together.

3 Copy the forward strand A protein called a DNA polymerase reads the first strand and starts building a copy on it.

4 Copy the backward strand DNA polymerases can only move in one direction, so they have to copy the second strand in small chunks

5 Join the fragments Proteins called DNA ligases heal the gaps between the copied pieces on the second strand.

Four vital molecules

DNA's organic bases make up the most important code in the universe

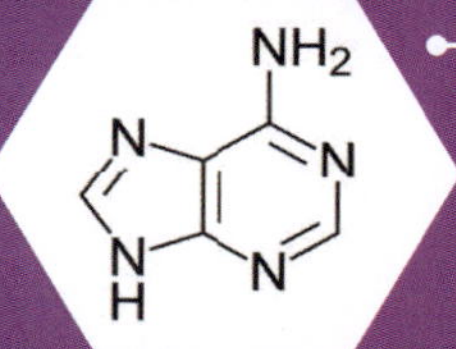

Adenine Adenine, A for short, is a purine base. This will always pair with thymine.

Cytosine Cytosine, C for short, is a pyrimidine base. It always pairs with guanine.

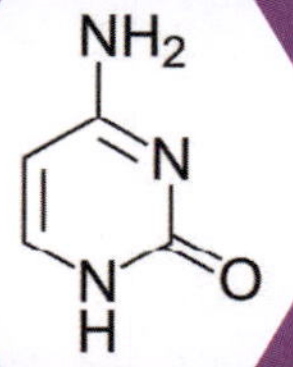

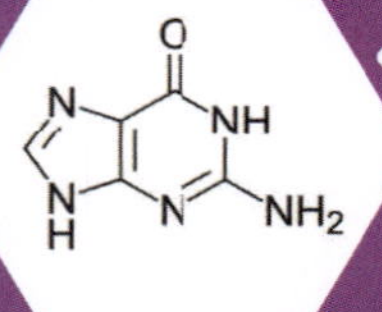

Guanine Guanine, G for short, is a purine base. It always pairs with cytosine.

Thymine Thymine, T for short, is a pyrimidine base. It always pairs with adenine.

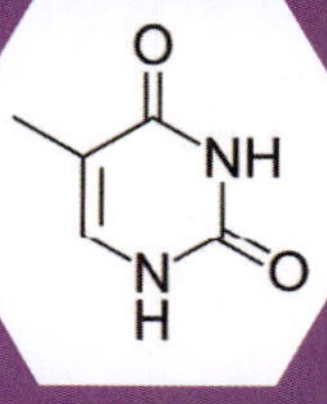

What is RNA?

DNA is not the only genetic molecule in nature; organisms also make strands of genetic material called RNA. Chemically the two molecules are quite similar, but they have very different roles to play. DNA forms stable double strands that carry the entire genome safely from one generation to the next. RNA forms unstable single strands that carry smaller amounts of genetic data for shorter amounts of time. Cells use RNA to make temporary copies of their genes. These copies act as templates for the molecular machines that make proteins. The machines read through the code three bases at a time, adding the corresponding amino acid to a growing protein string.

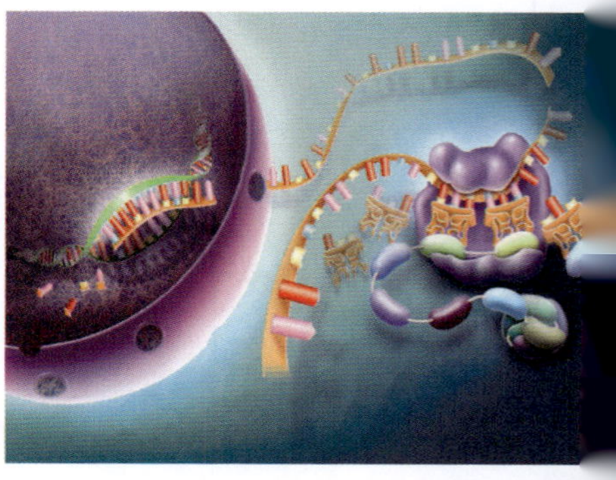

Cells copy their genes into RNA and use it as template to make protein

Twisted structure

How does two metres of DNA fit into a cell smaller than a grain of sand?

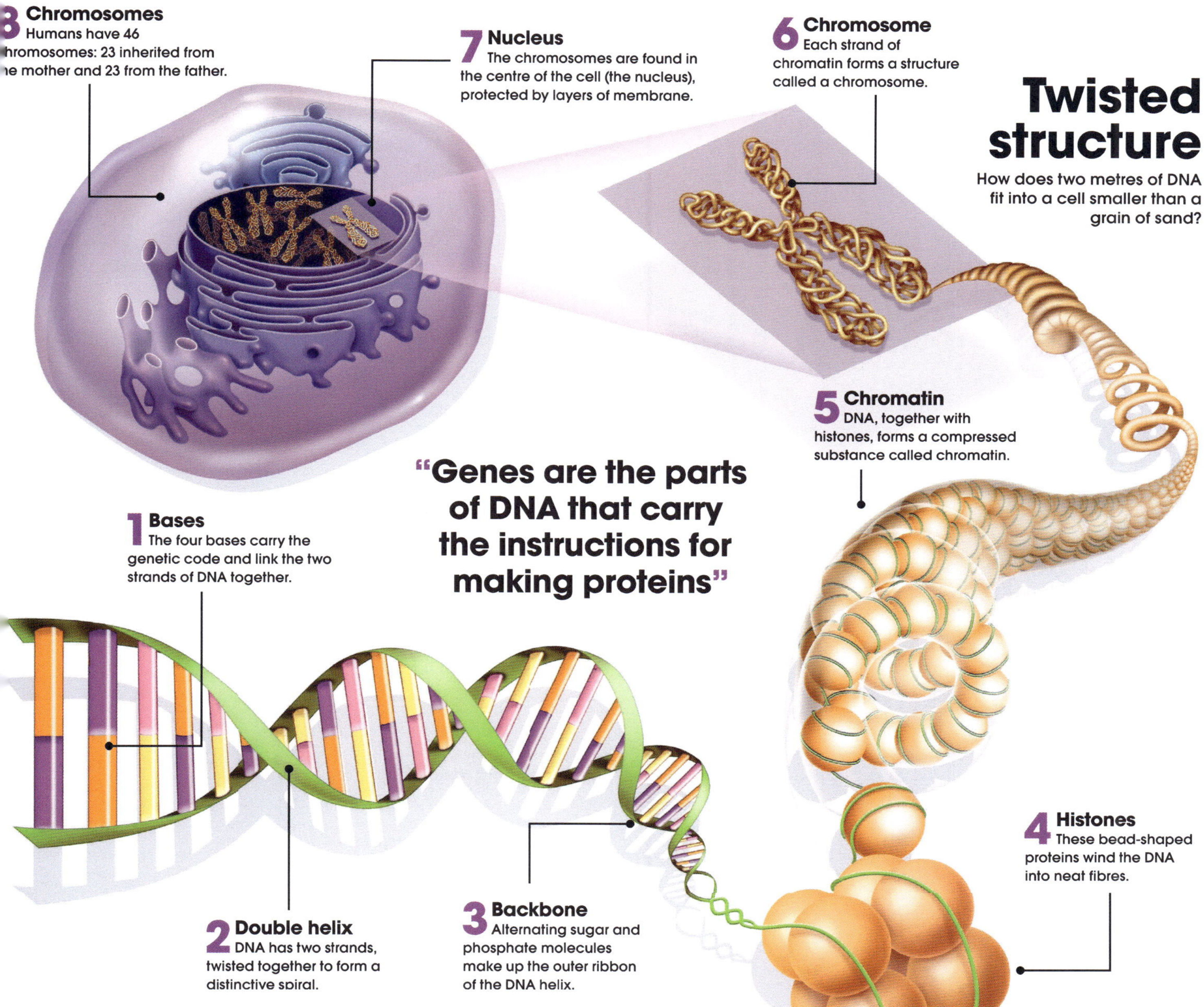

extremely picky about the bonds they will n. Adenine will only pair with thymine, and nine will only pair with cytosine. This means t if one strand carries the sequence ATGC, the er has to have the sequence TACG.

his strict pairing behaviour is what enables ll to easily make copies of its genetic code. mply unzips the two strands and then uses n both as templates to assemble two new nds, copying the matching nucleotides into empty slots.

he most important role for DNA is to carry rmation from one generation to the next. information is written in code, using the bases as chemical letters. Scientists have to decode the meaning of the entire human genome, but they have cracked the code that makes up our genes.

Genes are the parts of DNA that carry the instructions for making proteins: the molecules that build our bodies and do the hard biochemical work that keeps us alive.

The building blocks of proteins are amino acids, and there are 20 different types of amino acid to choose from. The order of amino acids in a protein dictates how the protein folds up to form a 3D structure and how it behaves inside the body.

The instructions for building proteins are written into our genes as strings of three-letter words called codons. It is possible to make 64 three-letter words using the four DNA bases. This means that, just like in English, some genetic words have the same meaning.

The codon ATG means 'start', signalling that the instructions to build a protein begin here. The codons TAA, TAG and TGA mean 'stop', signalling that the protein ends here. The rest correspond to different amino acids. Cells copy the genes they want to use and read through the codons one by one, selecting the matching amino acid and connecting it to a growing string. When they reach a stop codon, they know that the string is complete, and the protein is ready to fold into its 3D shape.

Chemical bonds

Bonding between chemicals is a simple matter of attraction

To understand the types of chemical bonding, we should first consider what the atoms of our universe are made of. To do this, we can use a simplified analogy of atomic structure by picturing an atom as a star being orbited by planets. The nucleus is the star, which is formed of positively charged protons and neutrons that have no charge. Negatively charged electrons are the orbiting planets.

In an atom's native form the number of electrons and protons are equal, which gives the atom a neutral charge overall. But other atoms are also attracted to these electrons, and when they're shared or stolen, a chemical bond is the result.

Types of bonding

How chemicals are bound together is largely dependent on how they share their electrons

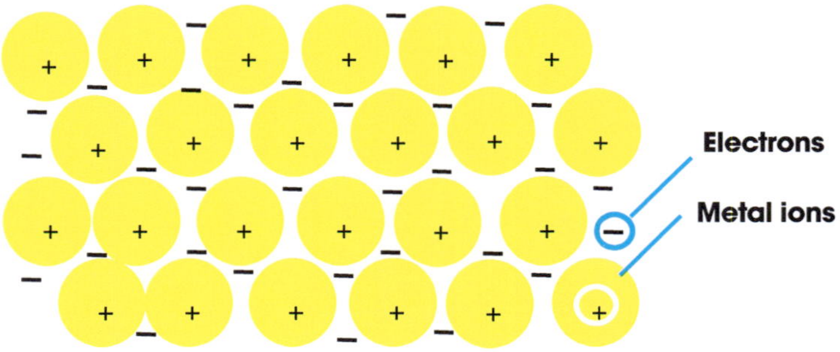

Metallic bonding

This type of bonding is strong and allows for the conduction of heat and electricity. Metals achieve this by universally sharing their electrons, becoming positive ions in a 'sea' of fluid electrons.

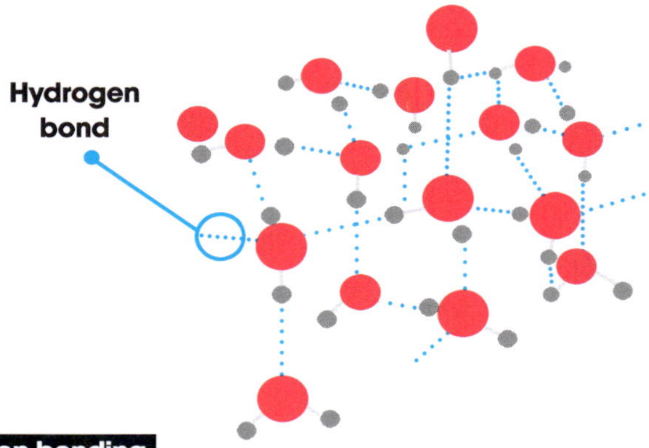

Hydrogen bonding

Hydrogen atoms carry a partial positive charge when part of a molecule, and form weak bonds with atoms of high electronegativity, which carry a partial negative charge. Many molecules can be bound via hydrogen bonding.

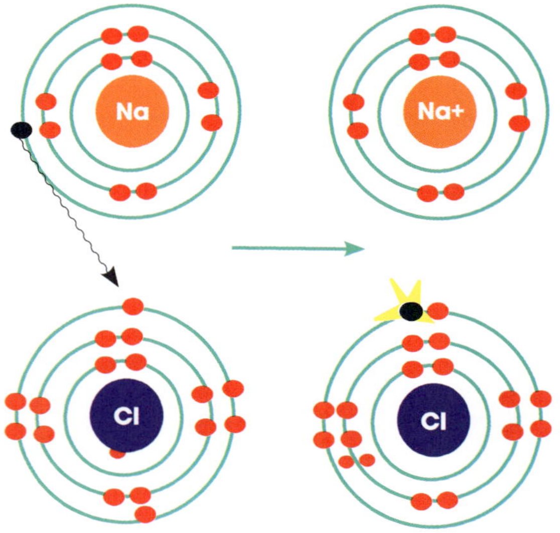

Ionic bonding

When an atom is less attracted to its outermost electrons, and another is greatly attracted to them, the electrons can be donated. This forms two ions – one positively charged and one negatively charged – that are attracted to each other.

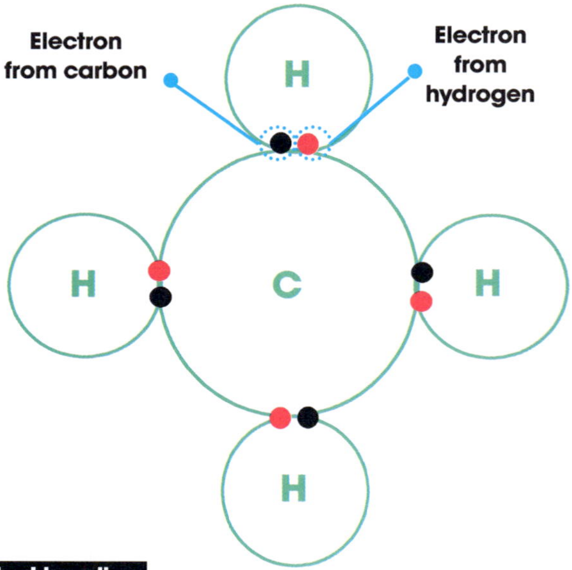

Covalent bonding

These bonds occur in nonmetal atoms that share electrons. When both atoms are quite strongly attracted to their neighbour's electrons, they can both make use of them, and will become bound together as molecules in the process.

Breaking down gas chromatography

Need to split a substance into its base components? Then you need to get hold of a gas chromatograph...

Chromatography is the process of separating a mixture into its core components for analysis or reuse. Gas chromatography uses an inert gas as a
rying mechanism within that separation, with
nples introduced to the gas and then filtered
ough a solid distillation column.
Key to the process is the retardation (slowing)
he mixture's elements in the column, which is
ieved by filling the latter with a finely divided
rt solid coated with a thin layer of reactive
id or polymer. The stream of gas passes the
ted solid and, while the inert gas – by its
y nature – won't react with it, the additional
ments it is carrying will.
ach component doesn't react with the
d simultaneously, however. This means
sample breaks down over time, with
h element leaving the column at a certain
nt depending on its composition. As such,
ixture of various substances can be split
its constituent parts, allowing each to be
ntified, studied and/or filtered.

"The sample breaks down over time, with each element aving the column at a certain point"

Gas chromatography step-by-step

Follow the stages of this chemical process up close with our handy guide

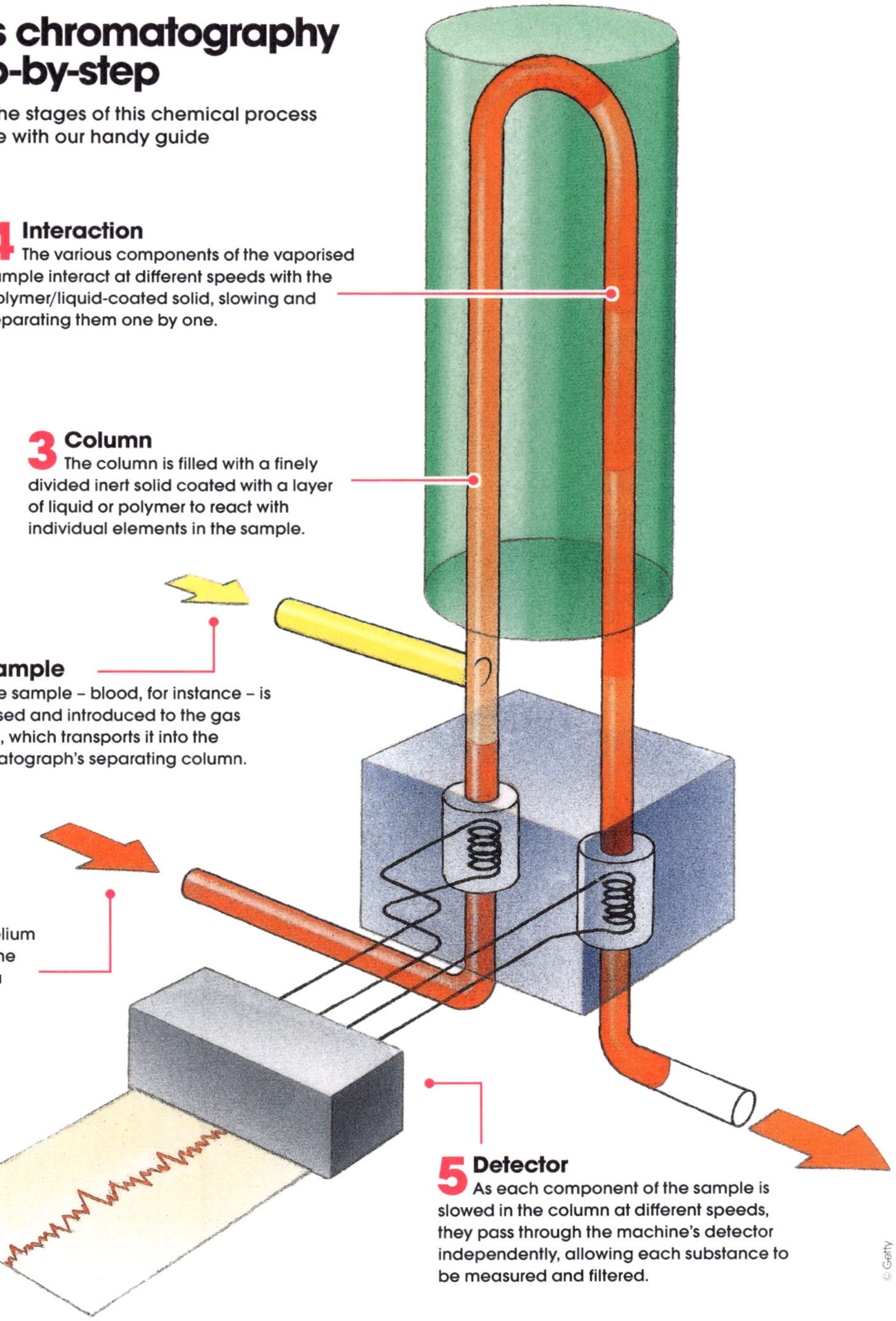

4 Interaction
The various components of the vaporised sample interact at different speeds with the polymer/liquid-coated solid, slowing and separating them one by one.

3 Column
The column is filled with a finely divided inert solid coated with a layer of liquid or polymer to react with individual elements in the sample.

2 Sample
The sample – blood, for instance – is vaporised and introduced to the gas stream, which transports it into the chromatograph's separating column.

1 Inert
A stream of inert gas – eg helium or argon – is pumped through the machine in a constant flow as a carrying mechanism.

5 Detector
As each component of the sample is slowed in the column at different speeds, they pass through the machine's detector independently, allowing each substance to be measured and filtered.

© Getty

The magic of carbon chemistry

Discover how this protean wonder-element can make more than 9 million different chemical compounds

Carbon is so versatile and so important that the Royal Society of Chemistry call it the 'King of the elements'. It's everywhere, and we use it for everything.

Carbon sits at position six on the periodic table. It has six protons, six neutrons and six electrons. Two of those electrons fill up an electron shell close to the nucleus, while the other four sit in a half-empty shell around the outside. These outer electrons, known as the valence electrons, are the key to carbon's incredible properties.

The electrons in an atom's outer shell are the ones that take part in chemical bonds. How many bonds depends on how much room there is in the shell, and the amazing thing about carbon is that it's got space for four. This means that it acts like a plumbing cross-piece, becoming a four-way connector that links different atoms together. It can build straight chains, chains with branches, and chains joined end to end to make loops. And it doesn't just bond to itself. The edges of carbon chains can terminate in tiny hydrogen atoms, making them slippery like oils. Or they can connect to other groups of elements, called 'functional groups', each of which has different chemical properties.

Stronger than steel

A stretched carbon fibre is twice as strong as steel but only a quarter of the density. Woven into mats and embedded in epoxy resin, this tough, flexible material is stiff, strong under tension and resistant to chemicals. It's resilient enough to build lightweight bikes, boats, cars, planes and artificial limbs, and it's all down to the way the layers of carbon slot together.

Individual layers of graphite are some of the strongest structures in the known universe, but they're slippery. Each honeycomb sheet of carbon atoms is strong on its own, but the stacks slide around on top of each other forming flakes and faults. In carbon fibres, the arrangement is different. The sheets of graphene aren't smooth. They form long, thin ribbons that intertwine in random folds, locking together to form tough, lightweight fibres that don't tear apart.

Carbon fibre is five times stronger than steel, and it's also lighter and more flexible than steel

Carbon works like elemental scaffolding, and the shapes, sizes and chemistry of the molecules it creates vary enormously.

Another important feature of carbon is that, although the bonds it makes are strong, they're not unbreakable. Carbon-based molecules are stable enough that they won't fall apart, but, given the right amount of energy, they can be reworked into something new. This makes carbon the perfect backbone for the molecules of life. Fats, carbohydrates, proteins and nucleic acids (DNA and RNA) are all built around carbon.

There are 100 million billion metric tons of carbon on Earth today, and it's all been around longer than the Solar System. Our Sun is the grandchild of the very first stars in the universe. It was born about 4.6 billion years ago out of a cloud of dust and gas created by a supernova explosion. That dust and gas contained every of atom carbon on our planet, and more.

Carbon reacts with oxygen to make the infrared-trapping gas, carbon dioxide. Too much of this can cause a runaway greenhouse effect that makes planets impossible to inhabit. Just look at Venus – its surface temperature is 462 degrees Celsius!

Luckily, Earth didn't become a greenhouse, it became an incubator for carbon-based life. And in a strange way, life and Earth have been working together to keep it that way ever since. Carbon is Earth's thermostat.

Human activities release 100 times more carbon into the atmosphere than volcanoes

Earth has a slow carbon cycle on a 100-milli year rotation. Carbon dioxide in the atmospher reacts with water to make carbonic acid, whicl eats away at the rocks. This releases trapped calcium. Shell-making organisms in Earth's oceans combine that calcium with carbon-containing ions to make calcium carbonate. When they die and drop to the sea floor, their remains become limestone, locking carbon aw underground. Slowly, this carbon returns to th atmosphere as Earth's crust shifts and shakes. Volcanic activity melts the rocks and releases

"Carbon is a four-way connector that links different atoms together"

Kinds of carbon

This versatile element comes in a variety of forms called allotropes

Amorphous

Prehistoric

A disorderly arrangement of carbon atoms is common in coal. It contains pockets of graphite and diamond, with properties in between.

It's hard to believe that coal and diamonds are made from the same element

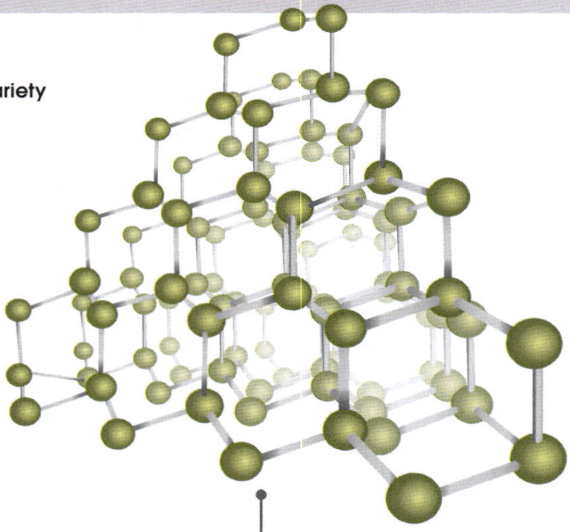

Diamond

Prehistoric

The carbons in a diamond form four bonds each, making tetrahedrons that link together to make 3D rings of six atoms.

Graphite

Prehistoric

Graphite contains slippery layers of carbon atoms arranged sheets of hexagons. Electrons move freely between them

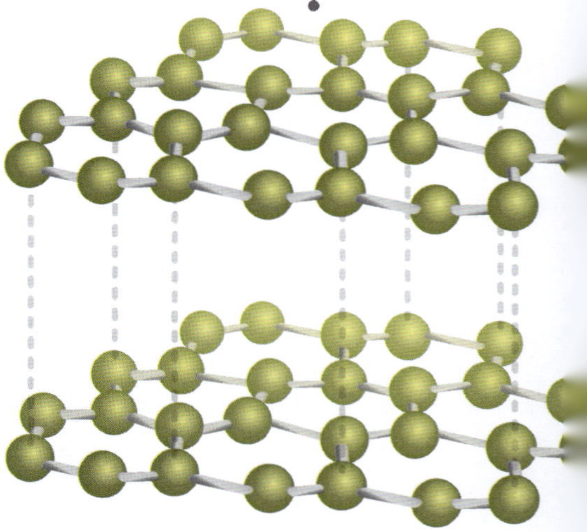

his aerial photo shows a
oal mine in action

Radiocarbon dating uses the decay of carbon-14 to assess the age of organic material

Buckminsterfullerene
1985
The original buckyball has 60 carbon atoms and is round in shape. They slip past each other and melt easily.

Nanotube
1991
These hollow cylinders look like rolled up graphene sheets. They resist stretching and conduct electricity.

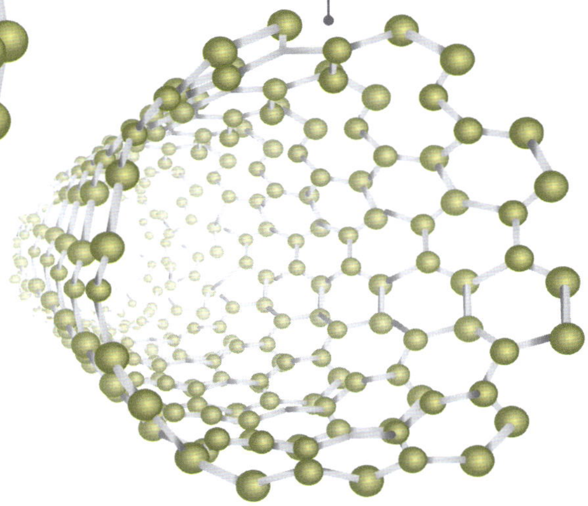

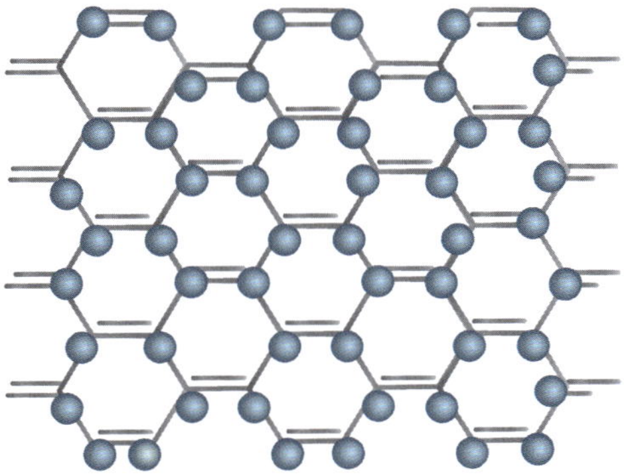

Graphene
2004
As the world's thinnest material it is 1 million times thinner than the diameter of a human hair. It is strong, lightweight and flexible – properties that make it useful in a range of industries.

Reading the carbon clock

High-energy cosmic rays help us to work out when ancient plants and animals died

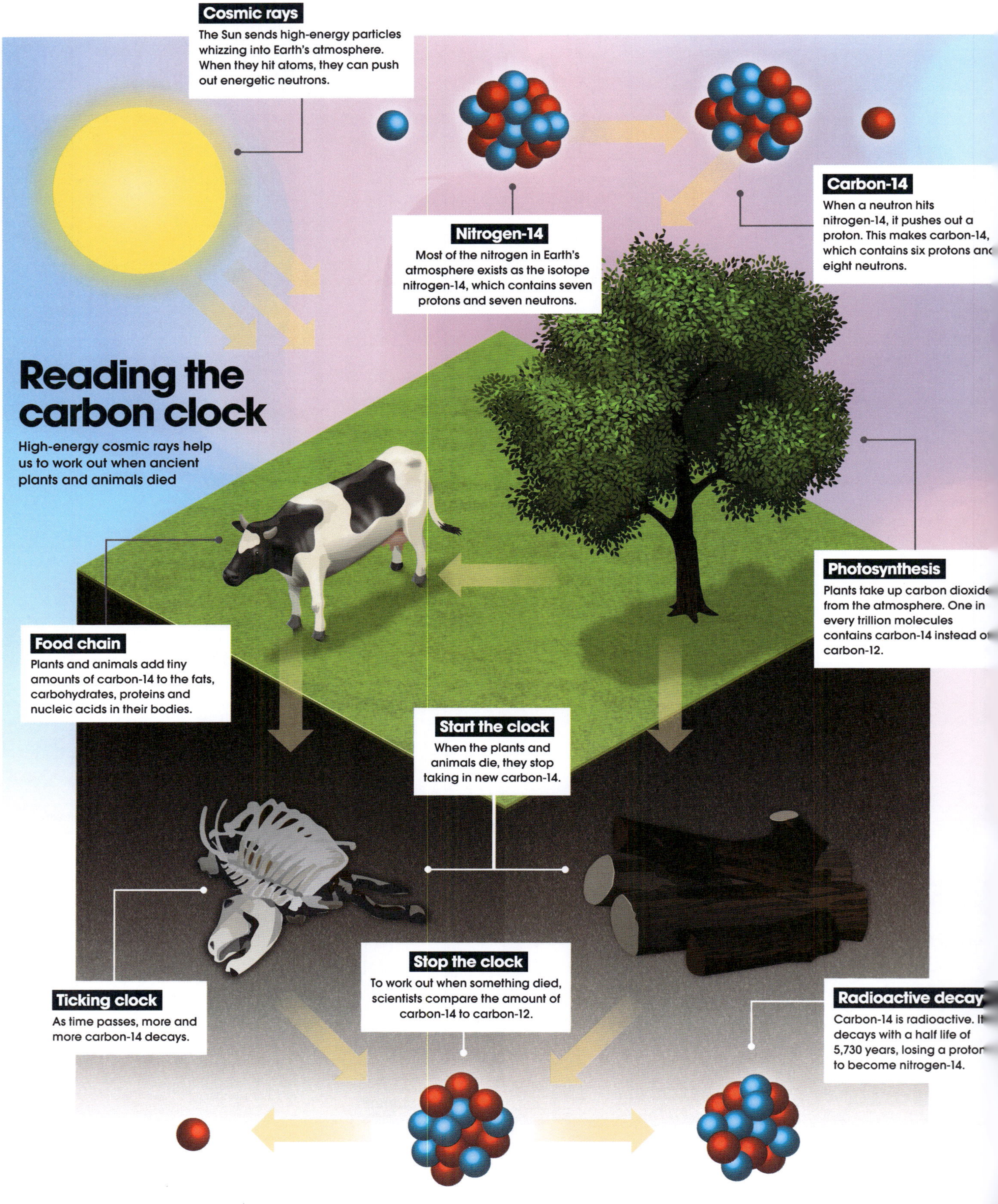

5 Facts
Carbon: the basic building block of life

1 You are 18.5% carbon
Almost all of your body is made up of just four elements. You're 65% oxygen, 18.5% carbon, almost 10% hydrogen and 3% nitrogen.

2 Pencils beat diamonds
Graphene is so strong that you can stretch it by around 25% before it snaps, which makes it stronger than both steel and diamond.

3 Carbon kills small stars
Helium fusion makes carbon, but the Sun is too small to burn it, so when the Sun runs out of helium, it will gradually grow cold and die.

4 Soot causes cancer
In 1775, doctors noticed a strange kind of skin cancer that only affected chimney sweeps. Contaminated soot from incomplete combustion was triggering the disease.

5 Carbon strengthens tyres
Car tyres get their colour from carbon, but it's not just there for show. It makes the tyres strong and protects them from the Sun.

Every living thing on Earth depends upon carbon chemistry

Volcanoes turn rocks into carbon dioxide, completing the carbon cycle

carbon back into the air at a rate of around 130 to 380 million metric tons every year.

Layered over the slow cycle is a fast carbon cycle, governed by plants. They take in carbon dioxide and use energy from sunlight to trap carbon in organic molecules. These molecules then pass through the food chain, becoming foliage and flesh before returning to the air as carbon dioxide.

Carbon has been so fundamental to life on our planet that its very existence is now inextricably linked to living things. Around 80 per cent of Earth's rocky carbon is limestone, made from the shells and skeletons of sea creatures. Most of the rest exists as fossil fuels, coals, oils and gases, made from the compressed remains of plants and animals.

The story of carbon is written into the very fabric of our planet, but it's not finished yet… Humans are currently releasing around 100 times more carbon into the atmosphere than volcanoes, and it's seriously messing with Earth's thermostat. Carbon dioxide levels were 280 parts per million 150 years ago, and now they're 412 parts per million.

This incredible element made our planet what it is today. Now, the race is on to find innovative ways to work with carbon to ensure our future.

Buckyballs in space

We used to think that molecules with more than 12 atoms wouldn't be able to survive in interstellar space, but we were wrong. In 2019, the Hubble Space Telescope confirmed that there are electric buckyballs floating between the stars. These giant molecules contain 60 carbon atoms each, and the telescope revealed that they're ionised. The energy from nearby stars has pulled away one of their electrons, leaving them with a positive charge. Finding these chunky molecules in outer space could help to answer some puzzling astronomical questions. Scientists have spotted more than 400 strange patterns of light absorption called Diffuse Interstellar Bands, or DIBs, made by unknown molecules in outer space. It could be that complex carbon chemicals are the cause.

Electric buckyballs have been spotted between the stars

The sulphur cycle

Always mixing and mingling, sulphur is an element that really likes to get around

The sulphur cycle is one of many biochemical processes where a chemical element or compound moves through the biotic and abiotic compartments of the Earth, changing its chemical form along the way. As with the carbon and nitrogen cycles, sulphur moves between the biosphere, atmosphere, hydrosphere and lithosphere (the rigid outer layer of the Earth). In biology, the water, oxygen, nitrogen, carbon, phosphorus and sulphur cycles are of particular interest because they are integral to the cycle of life.

Sulphur, present in the amino acids cysteine and methionine as well as the vitamin thiamine, is a vital part of all organic material. Plants get their supply from microorganisms in the soil and water, which convert it into usable organic forms. Animals acquire sulphur by consuming plants and one another. Both plants and animals release sulphur back into the ground and water as they die and are themselves broken down by microorganisms. This part of the cycle can form its own loop in both terrestrial and aquatic environments, as sulphur is consumed by plants and animals and released again through decomposition.

Elemental sulphur is found around volcanoes and geothermal vents, and when volcanoes erupt, massive quantities of sulphur, mostly in the form of sulphur dioxide, can be propelled into the atmosphere. Weathering of rocks and the production of volatile sulphur compounds in the ocean can also both lead to the release of sulphur. Increasingly, atmospheric sulphur is a result of human activity, such as the burning of fossil fuels.

Once in the air, the sulphur dioxide reacts with oxygen and water to form sulphate salts and sulphuric acid. These dissolve well in water and may return to Earth's surface via both wet and dry deposition. Not all the sulphur is getting busy; there are also vast reservoirs in the planet's crust as well as in oceanic sediments.

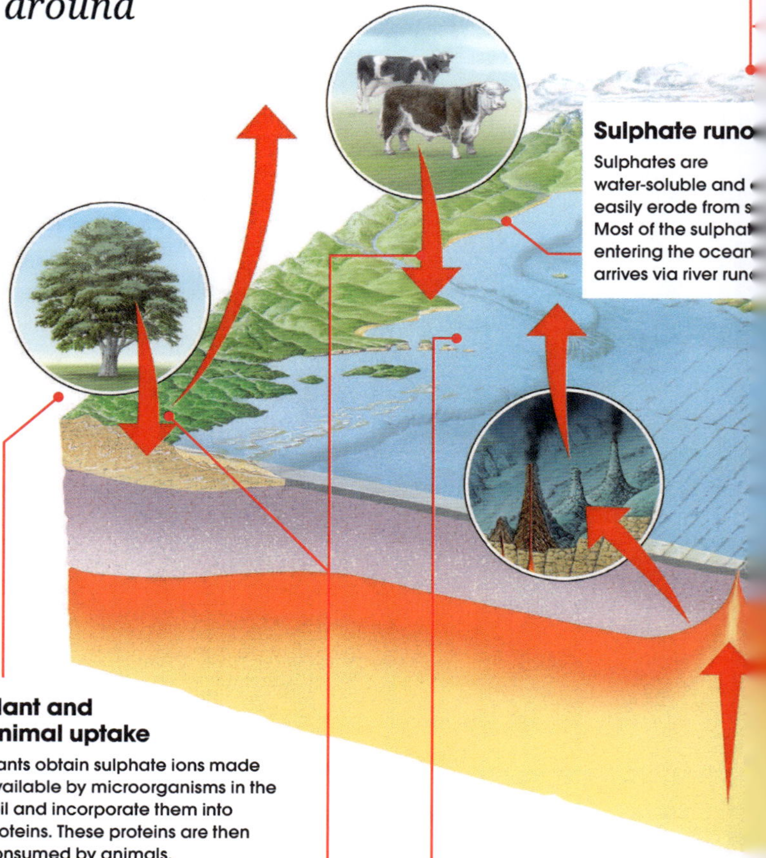

Atmospheric sulphur
Once in the atmosphere, some sulphur aerosols can remain for years, reflecting the Sun's energ back into space and lowering surface temperatures many mil away. The eruption of Mount Tambora in Indonesia is though to have caused the 'year withou summer' reported in Europe an North America in 1816.

Sulphate runo
Sulphates are water-soluble and easily erode from s Most of the sulpha entering the ocean arrives via river run

Plant and animal uptake
Plants obtain sulphate ions made available by microorganisms in the soil and incorporate them into proteins. These proteins are then consumed by animals.

Organic deposition
When biological material breaks down, sulphur is released by microbes in the form of hydrogen sulphide and sulphate salts, as well as organic sulphate esters and sulphonates.

Wet and dry deposition
The airborne deposition of sulphur compounds, whether sulphate sa or sulphuric acid, is the dominant cause of acidification in both terrestrial and coastal ecosystems

Sulphur and the climate

Human activities like burning fossil fuels and processing metals generate around 90 per cent of the sulphur dioxide in the atmosphere. This sulphur reacts with water to produce sulphuric acid and with other emission products to create sulphur salts. These new compounds fall back to Earth, often in the form of acid rain. This type of acid deposition can have catastrophic effects on natural communities, upsetting the chemical balance of waterways, killing fish and plant life. If particularly concentrated, acid rain can even damage buildings and cause chemical weathering.

However, the environmental impact of sulphur pollution isn't entirely negative; atmospheric sulphur contributes to cloud formation and absorbs ultraviolet light, somewhat offsetting the temperature increases caused by the greenhouse effect. In addition, when acid rain deposits sulphur in bodies of wetlands, the sulphur-consuming bacteria quickly out-compete methane-producing microbes, reducing the methane emissions (roughly 22 per cent of the human-induced greenhouse effect).

Burning fossil fuels accounts for a large proportion of the sul dioxide in the atmos

The cycle in action

Sulphur is ubiquitous on Earth but much like your average teenager, the behaviour of sulphur depends heavily on its companions. The element is both necessary for all life and potentially highly toxic, depending on the chemical compound. It moves through different compartments of the planet, taking a range of forms, with many and varied impacts.

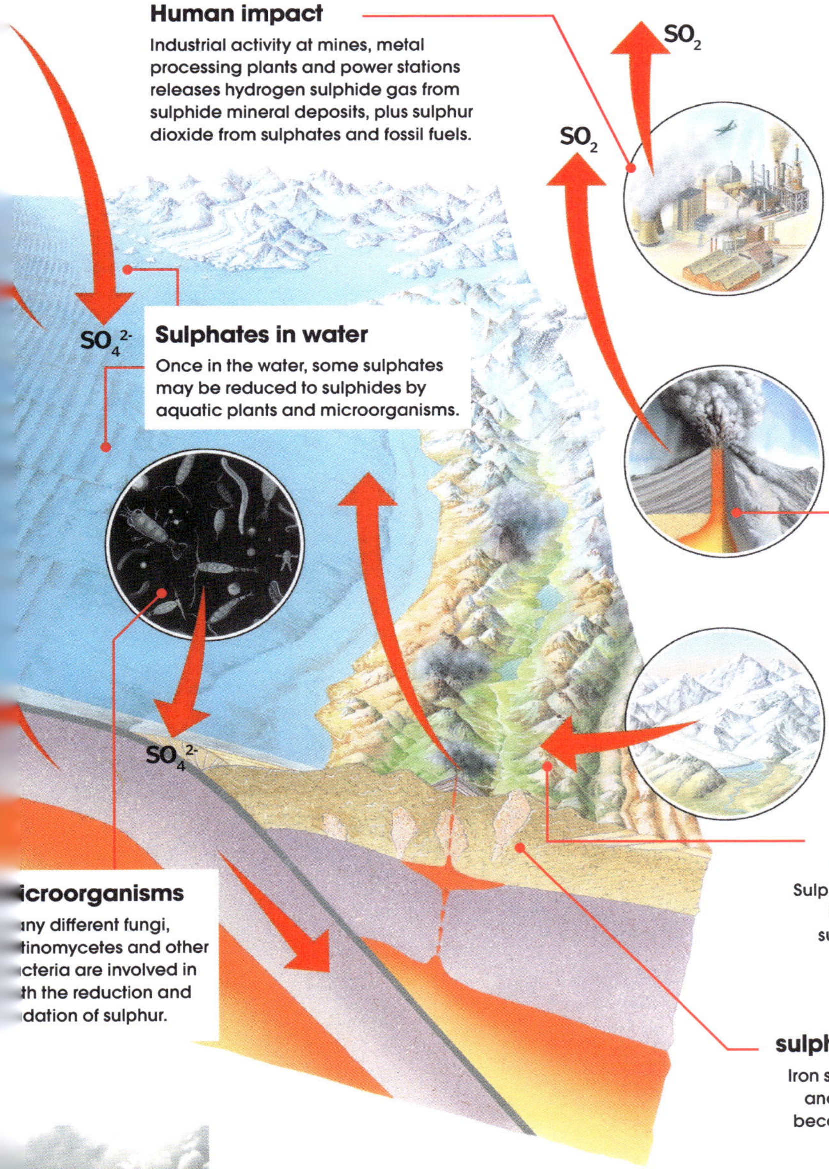

Human impact

Industrial activity at mines, metal processing plants and power stations releases hydrogen sulphide gas from sulphide mineral deposits, plus sulphur dioxide from sulphates and fossil fuels.

Sulphates in water

Once in the water, some sulphates may be reduced to sulphides by aquatic plants and microorganisms.

Release of sedimentary sulphur

Volcanic and industrial activity release hydrogen sulphide gas from sulphide mineral deposits, and sulphur dioxide from sulphates and fossil fuels.

icroorganisms

ıny different fungi, tinomycetes and other ıcteria are involved in th the reduction and ıdation of sulphur.

Deposition of sulphate minerals

Sulphates are also deposited in sediments as minerals, such as gypsum, a form of calcium sulphate.

Deposition of sulphides in sediments

Iron sulphide, known as pyrite, and other sulphide minerals become buried in sediments.

Large quantities of sulphur in its mineral form are found around volcanoes

Its yellow colour led some alchemists to try and re-create gold with sulphur

What is sulphur?

Sulphur is one of the most important and common elements on Earth. It exists in its pure form as a non-metallic solid and is also found in many organic and inorganic compounds. It can be found throughout the Earth's environment, from the soil, air and rocks all the way through to plants and animals.

Because of its bright yellow colour, sulphur was used by early alchemists in their attempts to synthesise gold. That didn't pan out, but people still found many useful applications for it, including making black gunpowder. Today sulphur and sulphur compounds are used in many consumer products such as matches and insecticides. Sulphur is also a common garden additive, bleaching agent and fruit preservative, and is an important industrial chemical in the form of sulphuric acid.

Early users mined elemental sulphur from volcanic deposits, but when the demand for sulphur outstripped supply towards the end of the 19th century, other sources had to be found. Advances in mining techniques enabled the extraction of sulphur from the large salt domes found along the Gulf Coast of the United States. Both volcanic and underground sulphur deposits still contribute to the global supply, but increasingly, industrial sulphur is obtained as a byproduct of natural gas and petroleum refinery processes.

How do noble gases work?

What makes this select bunch of chemical elements so 'noble'?

There are six naturally occurring noble gases found around our world and beyond. These are helium (He), neon (Ne), argon (Ar), krypton (Kr), xenon (Xe) and radon (Rn). Together they form Group 18 of the periodic table and are characterised by their lack of colour, smell, taste and flammability in their natural state.

Despite being historically referred to as rare and inert, noble gases – which were designated 'noble' due to their apparent reluctance to undergo a chemical reaction – are nothing of the sort. In fact, all of these gases are found in Earth's atmosphere and each is capable of being chemically active and producing compounds.

The majority of the noble gases – ie argon, krypton, neon and xenon – are formed via liquefaction and fractional distillation techniques, however helium is attained by separating it from natural gas and radon by isolating it from the radioactive decay of radium compounds.

As noble gases show extremely low chemical reactivity, while they are not inert, only a few hundred noble gas compounds have been formed to date, with xenon varieties making up the bulk. In theory, though, radon is more reactive than xenon, so should form chemical bonds more readily. However, its high radioactivity and short half-life are the key factors which prevent this.

There are many applications for noble gases (see the boxout for some notable examples). The most obvious and visible of these are illuminated signs, light bulbs and lamps, with xenon, argon and neon commonly used due to their lack of chemical reactivity. Using these gases helps to preserve filaments in light bulbs and grants distinctive colours when used in gas-discharge lamps – as demonstrated by the main image used here.

Where are noble gases used?

Arc lamps

A specialised type of gas-discharge lamp, arc lights pass electricity through a bulb full of ionised gas, such as xenon or argon. They're used in IMAX cinemas among other places.

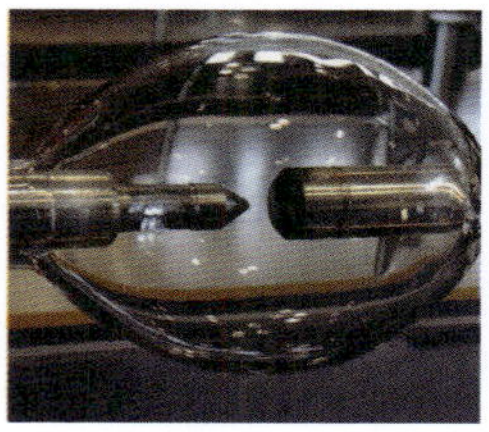

Blimps

Today, most blimps are filled with helium due to its lightness and incombustibility. Hydrogen was used originally but was phased out due to its high flammability.

MRI scanners

One of the most advanced pieces of medical equipment, magnetic resonance imaging scanners use liquid helium to cool the superconducting magnets inside.

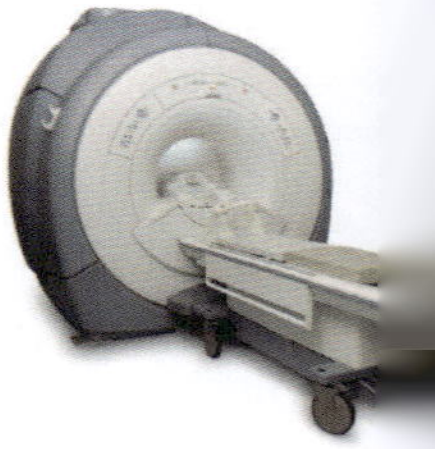

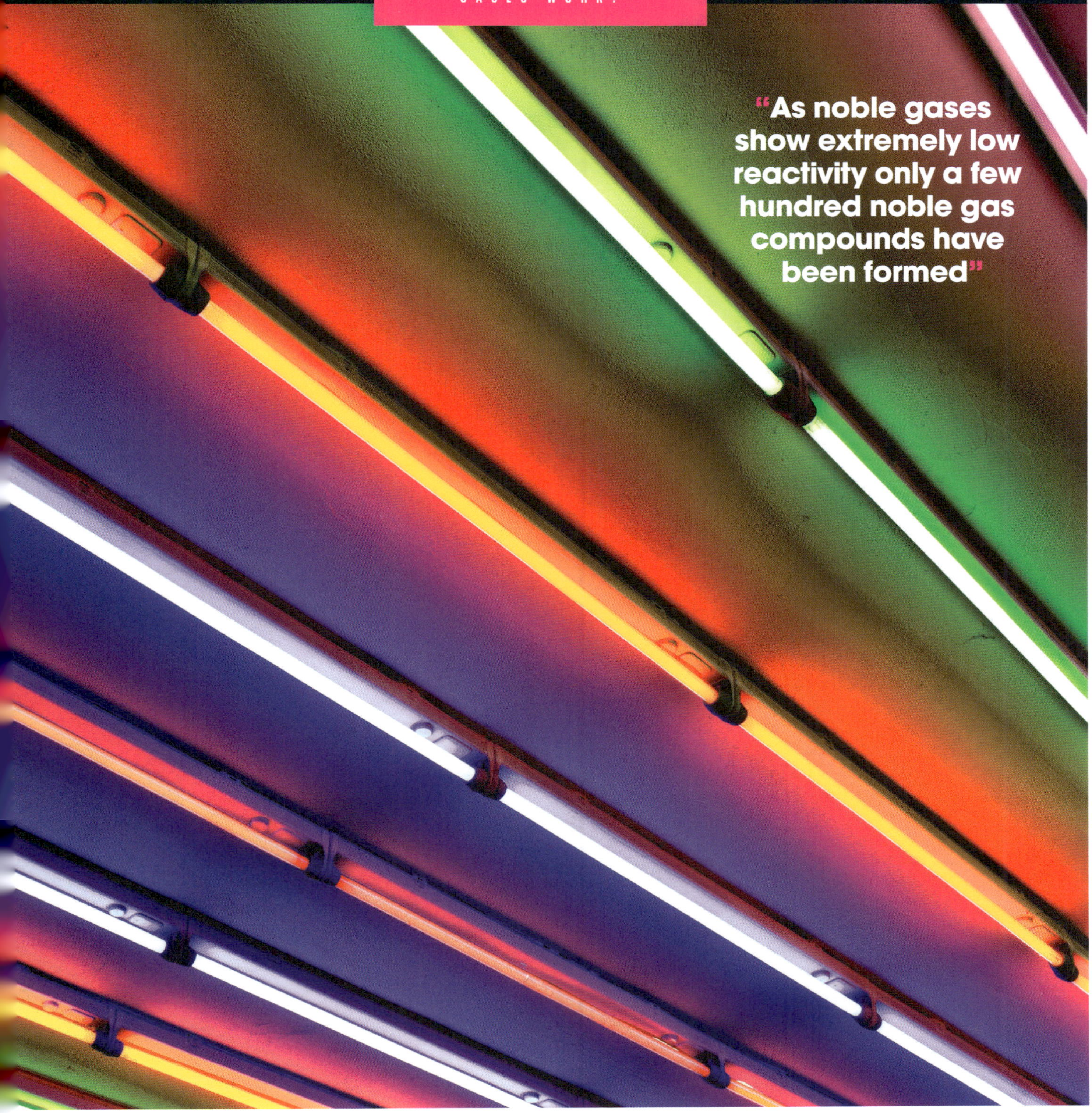

"As noble gases show extremely low reactivity only a few hundred noble gas compounds have been formed"

uminated signs

ny illuminated signs nd billboards utilise noble gases due to their ability to generate vibrant olours when ionised neon lights being a prime example.

Refrigerants

Due to their incredibly low boiling points – for instance, argon boils at -186 degrees Celsius (-302.8 degrees Fahrenheit) – the Group 18 gases are often used in cryogenics.

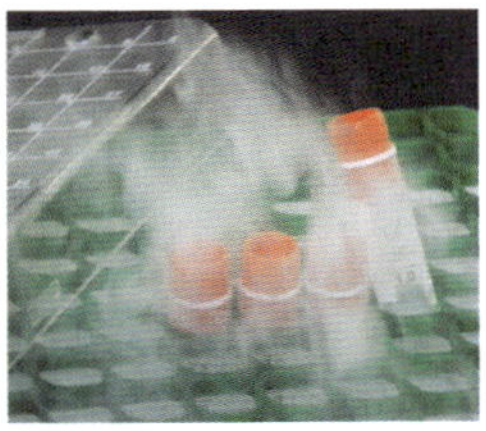

Radiotherapy

Despite the noble gas radon being highly radioactive and able to cause cancer, it can also be used as part of radiotherapy treatments to control or kill malignant cells.

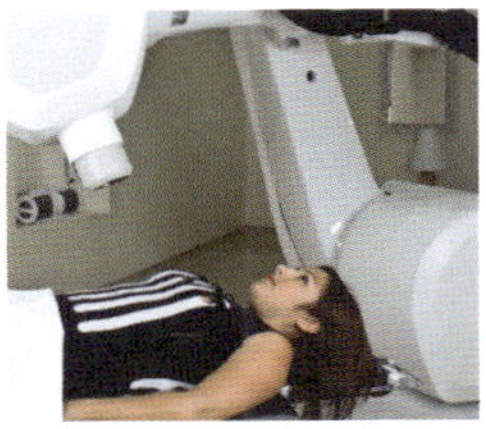

Making light of helium

Known best for making balloons float, this element's uses are surprisingly versatile

Floating in a party balloon, helium seems like harmless fun – but you also find it at the furthest extremes of heat and cold. Helium is the second-simplest chemical element: each atom is built from just two protons, two neutrons and two electrons. It can be created when four atoms of the very simplest element, hydrogen, which has one proton and one electron, fuse together. Hydrogen transforms like this in the hot, dense conditions we find in stars like the Sun. The hydrogen-to-helium nuclear fusion process creates the awesome power that takes the Sun's core temperature to around 15 million degrees Celsius.

Helium is the second most abundant element in the universe after hydrogen. But on Earth, it's rare. The combination of two protons, two neutrons and two electrons give it properties that mean it doesn't stick around for long, including in the lightness so prized for balloons. Being so lightweight, it can easily escape the planet's gravitational pull and drift out into space. It's also chemically very stable, so doesn't react easily with heavier substances that might keep it on Earth. That stability is what makes helium preferable for use in balloons to hydrogen, which burns easily.

What's even more useful about helium is that despite being born from the fiery heart of stars, it can be the coldest thing on Earth. Changing it from gas to liquid means cooling it below its boiling point of -268.9 degrees Celsius. This makes helium valuable for accessing properties that only show up when materials are this cold, such as using superconductor materials that can be very strong magnets, found in MRI scanners. For such a lightweight element, helium has a heavyweight impact when used like this.

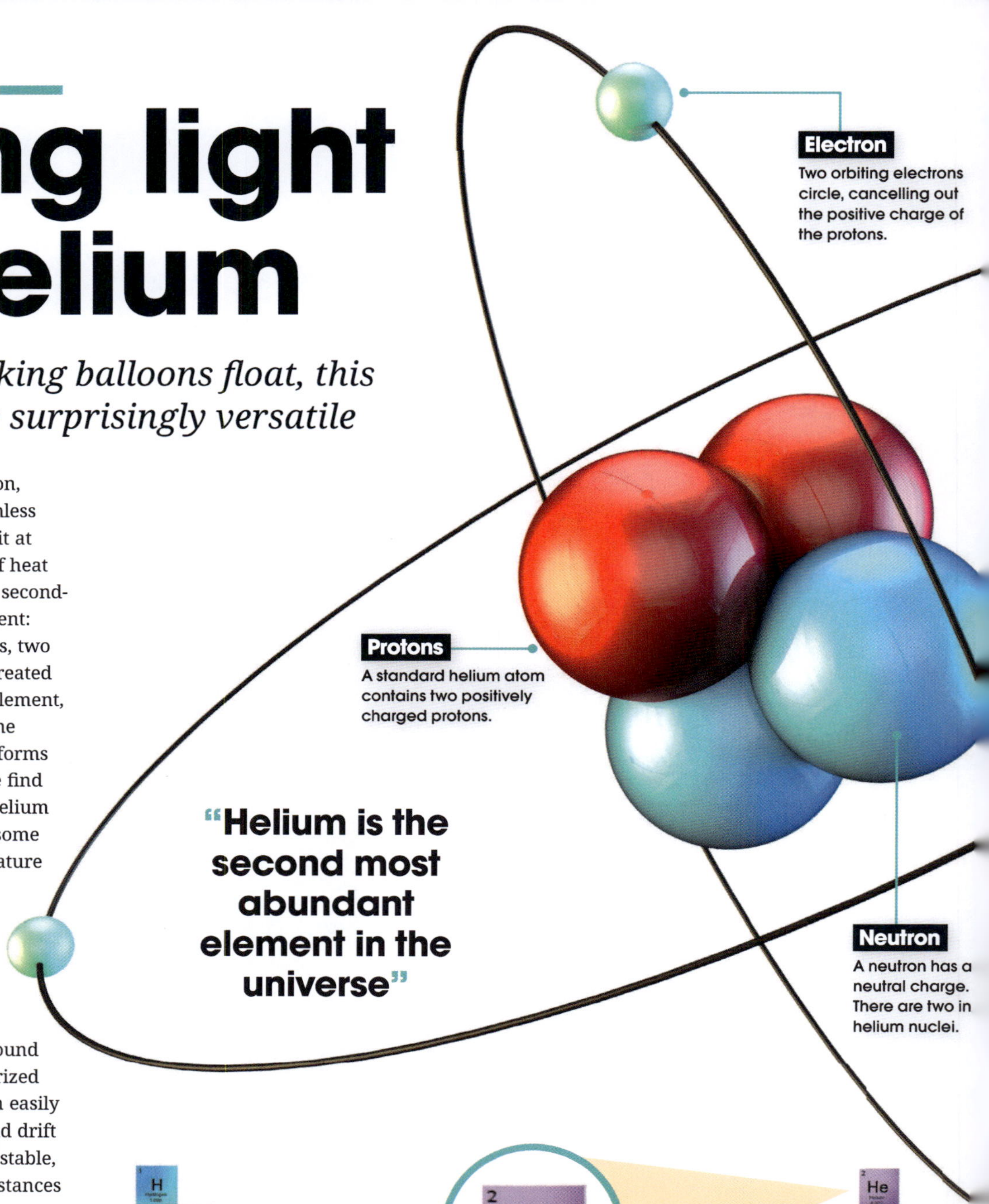

"Helium is the second most abundant element in the universe"

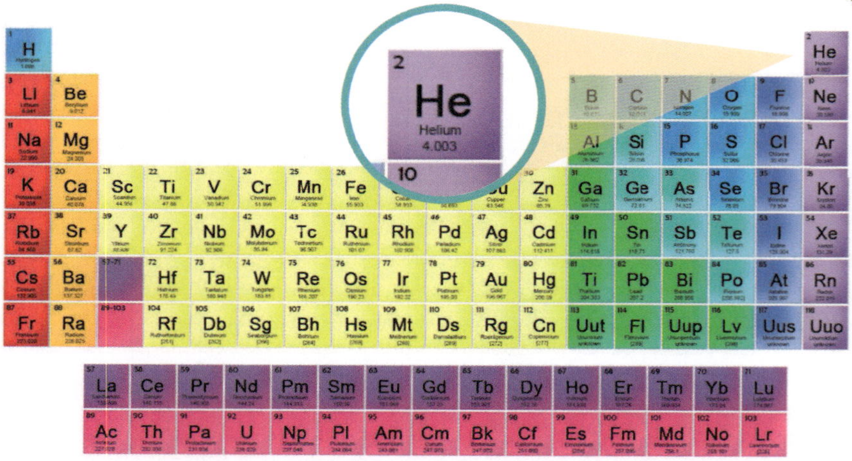

Helium is the first of the noble gases in the periodic table of elements

ignpost in Kansas marks the place
ere a huge source of gas that
ouldn't burn' was discovered in 1903

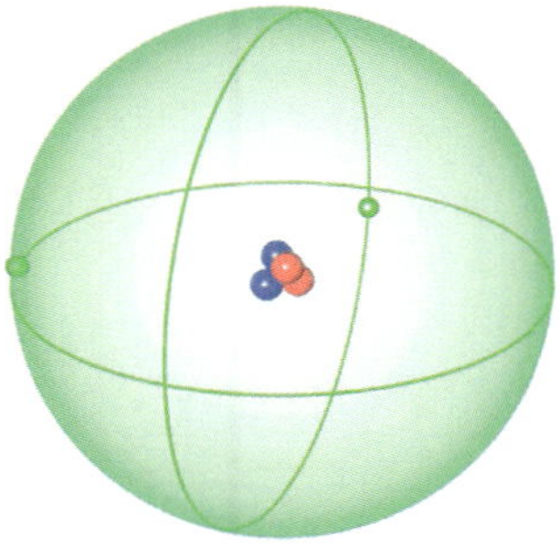

AR zone SCAN HERE

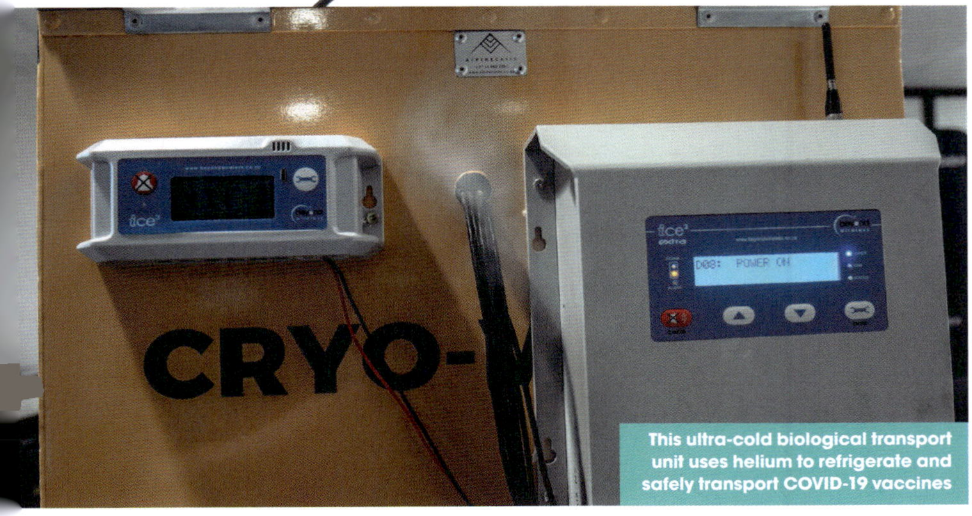

This ultra-cold biological transport unit uses helium to refrigerate and safely transport COVID-19 vaccines

A gap in the rainbow

Scientists in the 19th century found helium in sunlight using instruments called spectroscopes. These break up light like a rainbow, but much more precisely. Spectroscopes revealed 574 dark lines in the rainbow. In 1859, Gustav Kirchhoff and Robert Bunsen noticed that some lines exactly matched similar lines found when burning known chemical elements. They realised the lines revealed the Sun's chemical composition.

During a solar eclipse in August 1868, Pierre Janssen saw a line never previously seen. In October the same year, English astronomer Norman Lockyer also saw the line. Both decided it was a new element. Lockyer called it helium after the Greek word for the Sun, helios.

For decades, no one could find helium on Earth. Then, in 1895, William Ramsay detected helium when a mineral called cleveite reacted with acid. The radioactive element uranium had broken down to make helium, which was trapped in the rock but freed by the acid.

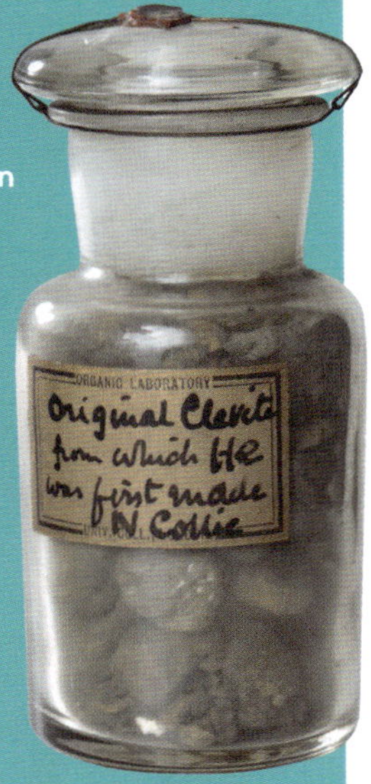

Uranium atoms in cleveite minerals break down and result in helium

Running out of gas

You might think we get helium by pulling it out of the air, but actually we suck it out of the ground. Underground stores form very slowly as atoms of large radioactive elements break down, releasing helium gas. Through the 20th century the US government was the world's largest helium supplier, having stockpiled it for use in military balloons. A former gas field in Texas holds enough helium for about 3 billion party balloons. The US government sold the Federal Helium Reserve in January 2024.

There is a feared helium shortage, with the American Chemical Society calling it an 'endangered element', and advising people don't buy helium party balloons. For now Russia is stepping in to fill the supply gap. However, it's still worth remembering that helium forms on Earth very slowly.

Ultracool, super-stable and buoyant

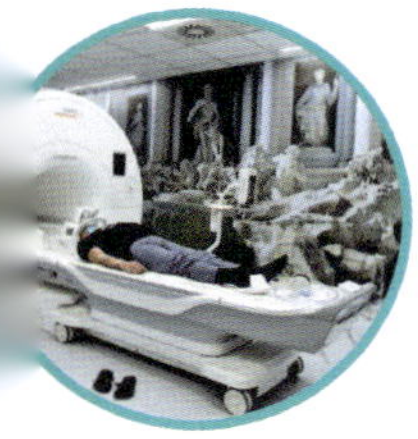

Seeing inside our bodies

iquid helium circulates
e magnetic resonance
ging (MRI) machines to
oduce scans that help
tors see inside patients.
e liquid helium cools
superconducting
bium-titanium magnets.
MRI machine makes the
gnetic field oscillate so
hydrogen atoms in the
ient's tissues give off a
tectable radio signal.

Keeping safe in space

Hydrogen's boiling point is -252.88 degrees Celsius, so helium is able to remain a gas even at the same low temperatures as liquid hydrogen, which is used as rocket fuel. Helium can therefore safely be used to clean out spacecraft engines. In its liquid form, helium is also used to cool down satellite instruments that might overheat.

Isolating chemicals

Because helium is so unreactive, scientists don't have to worry about it changing substances that it comes into contact with. That means it can safely help separate mixtures of chemicals using gas chromatography. It acts like the water that separates ink into different colours when printed paper gets wet.

Studying the upper atmosphere

Helium-filled weather balloons venture into Earth's upper atmosphere every day. At roughly midnight and noon Greenwich Mean Time (GMT), meteorologists release almost 900 balloons around the world. They can swell to five times their original size as the pressure outside the rising balloon decreases.

Hydrogen: the unifying element

The most common element in the universe is actually quite rare here on Earth

Hydrogen is the most important element there is. It is both dangerous and essential – just two of its amazing qualities. Hydrogen can be found in our bodies and many things around us. Also, hydrogen connects us all to the start of the universe!

The special qualities of hydrogen start from its building blocks. In its most familiar form, hydrogen is the simplest element. Its atoms have just one proton and one electron. Like most other elements, it has isotopes with extra neutrons. But hydrogen is the only element with named isotopes. It's called deuterium when it has one neutron and tritium when it has two. This is more than just a curious fact – our lives rely on hydrogen's isotopes.

That's because at the heart of the Sun, hydrogen atoms fuse together to make deuterium. One proton becomes a neutron in the process, releasing energy that becomes the heat and light we rely on. This is a similar process to the one that happened following the Big Bang, making hydrogen and leading to all the other elements. Most of the hydrogen in our bodies and everything else was made then, tying us to the start of the universe.

This simple atomic structure makes hydrog[illegible] much lighter than air. It enabled human fligh[illegible] in airships until another of hydrogen's key properties had its say: it combines so well with oxygen that it burns easily, infamously destroying an airship called the Hindenburg i[illegible] 1937. Yet as they burn, hydrogen and oxygen make water, which we need to stay alive.

Hydrogen reacts similarly well with many elements, making important substances. The materials hydrogen creates with carbon are t[illegible] building blocks of living things like us. With s[illegible] many important roles, hydrogen is truly the f[illegible] among the elements.

Liquid hydrogen, burned together with oxygen, is the most efficient rocket fuel

"Hydrogen connects us all to the start of the universe"

How we use hydrogen

cket fuel

n it is burned ther with oxygen, ogen is the most ient rocket fuel there has powered many e launches, ding NASA's Space tles. It's used as a d, meaning it must be very cold, which is a cult task to do.

Oil refining

The UK alone uses 100,000 tonnes of hydrogen a year to turn natural gas and oil into useful products. When they come out of the ground, gas and oil contain sulphur, which creates pollution when it's burned. Hydrogen helps remove the sulphur.

Making fertiliser

Farmers can produce much more food than in the past thanks to fertiliser. This gives plants the nitrogen they need to grow in the form of ammonia. Large factories make ammonia from hydrogen and nitrogen from the air, using the Haber-Bosch process.

Mining metals

A lot of the metals that we use come out of the ground locked up together with oxygen in rocks. Because hydrogen reacts so well with oxygen, it can pull it away from the rock. This leaves pure metal in a usable form, and the only by-product of this is water.

An elementary discovery

Henry Cavendish was the first to realise hydrogen was an element

Some credit for discovering hydrogen goes to Theophrastus Philippus Aureolus Bombastus von Hohenheim. Better known as Paracelsus, around 1520 he recorded that dropping metals into strong acids released bubbles of gas. In 1650, Théodore de Mayerne repeated the experiment, discovering that the gas burned easily. But it wasn't until 1755 that people realised different types of gases existed.

Soon after, the first person to work out that hydrogen was a unique substance was Henry Cavendish, who called it 'inflammable air'. At the time, some people thought that water and 'dephlogisticated air' – what we now call oxygen – were the only basic elements. In experiments from the 1760s and 1780s, Cavendish found that when inflammable air burned, it combined with dephlogisticated air to form water. That led French chemist Antoine Lavoisier to propose calling the gas hydrogen, as it is the gene, or creator, of hydro, which is Greek for water.

Born in the Big Bang

When he first thought it up, not everyone believed George Gamow's idea of how the elements were created. He suggested they formed by adding protons and neutrons together one at a time in the Big Bang. This happened when the universe was hot and dense, before it expanded and cooled rapidly. Working mainly with Ralph Alpher, Gamow saw the early universe as a highly compressed soup of neutrons. As the universe expanded, there would be room for neutrons to escape the soup and decay to make protons and electrons. From there hydrogen and deuterium might form, and then the other elements. Today this idea explains the large amount of hydrogen in the universe. However, other ideas were needed to explain how other elements were made.

All other elements were born from hydrogen

Acids and bases

Discover the differences between acids and bases, and find out why they act the way they do

It is widely known that lemons taste sour due to their acid content, soil needs the optimum pH level for plants to grow properly and acid rain can wipe out entire ecosystems. But what really makes one thing acidic and the other one basic (alkaline)? Why can they be so corrosive? And why does litmus paper turn different colours when dipped in acid or a base?

Acids and bases can be defined in terms of their concentration of hydrogen ions. Normally an atom of hydrogen consists of one proton and one electron, giving it a balanced electrical charge – protons being positively charged and electrons being negatively charged. Take away the electron and you are left with an ion of hydrogen, or a single proton, or 'H+', as it is often written. The thing about ions is they are very reactive, as they no longer have a balanced charge. They are constantly seeking ions of the opposite charge – an atom or molecule with an unequal number of electrons than protons, with which to react.

A strong acid has a high concentration of H+ ions and is defined by its ability to 'donate' hydrogen ions to a solution, whereas a base, also know as an alkali, has a much lower concentration of H+ ions and is defined by its ability to 'accept' hydrogen ions in a solution. Therefore, acids mixed with bases become less acidic and bases mixed with acids become less basic, or less alkaline.

"A strong acid has a high concentration of H+ ions"

Certain concentrated bases can attack living tissue and cause severe burns due to the ions reacting with the skin. However, the process of bases reacting with the skin, and other materials, is different to that of acids. That's why we call some concentrated acids 'corrosive', and reactive concentrated bases are 'caustic'.

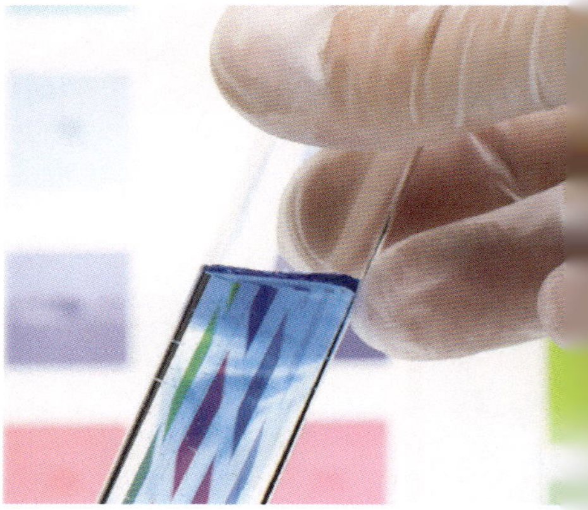

The power of hydrogen

The letters pH stand for 'power of hydrogen', as the scale refers to the concentration of hydrogen (H+) ions in the solution. It measures the acidity or basicity of a solution, with pH values ranging from 0-14, 0 being really acidic and 14 being really basic. A substance in the middle of the scale with a pH of 7 is classed as neutral, as it contains equal numbers of oppositely charged ions.

Acid

A compound which 'donates' hydrogen ions when placed in an aqueous solution. The higher the concentration of hydrogen ions released, the stronger the acid.

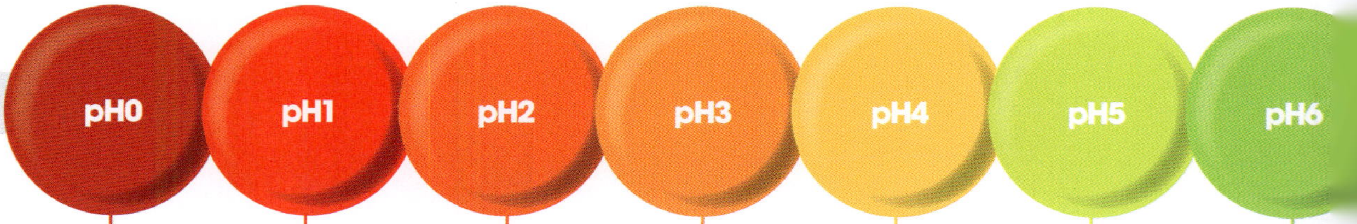

Hydrofluoric acid
A highly corrosive substance, which as a gas is a severe poison and acts as a catalyst in oil refining.

Suphuric acid
Used in the production of fertilisers, this strong acid is a chemical found in acid rain.

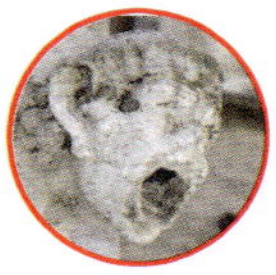

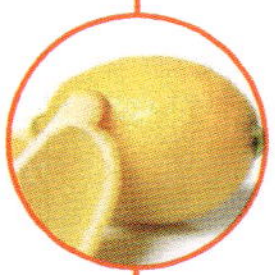

Lemon juice
Lemon juice is about 5% citric acid – a weak organic acid that gives lemons their sour taste.

Cola
Cola contains phosphoric acid, something that has been linked to the lowering of bone density in studies.

Beer
Various acids are formed during the fermentation process in beer production. The addition of CO2 also causes the pH to lower slightly.

Black coffee
Levels of acids found in coffee can be affected by the altitude at which the coffee was grown and the minerals present in the soil.

Cow's m
Milk goes so
time due to
bacteria pro
lactic acid
of a proc
of ferment

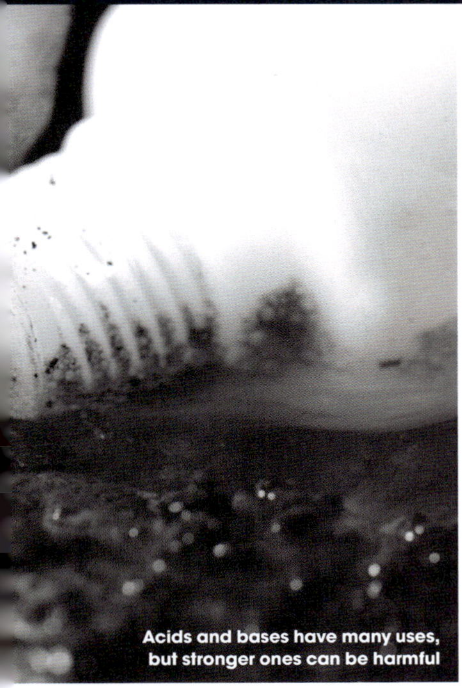

Acids and bases have many uses, but stronger ones can be harmful

he litmus test

can test the acidity or alkalinity of a substance ng litmus paper. Litmus paper is that which has en treated with a mixture of 10-15 natural dyes ained from lichens. The dyes work as icators, whereby upon exposure to acids (a pH s than 7) the paper turns red and upon osure to bases (a pH more than 7) the paper ns blue. When the pH is neutral (pH equal to 7), dyes cause the paper to turn purple. ed cabbage juice can also be used to tinguish between acids and bases, as it ntains a natural pH indicator called 'flavin'. on exposure to acid, flavin turns a red colour, tral solutions appear a purple colour and the sic solutions result in a greenish-yellow colour.

Neutralisation

A neutralisation reaction is the combination of an acid and base that results in a salt and, usually, water. In strong bases and acids, neutralisation is the result of the exchange of hydrogen and hydroxide ions, H+ and H- respectively, which produces water. With weak acids and bases, neutralisation is simply the transfer of protons from an acid to a base. The production of water, with a neutral pH of 7, indicates the neutralisation of the acid and base, while the resultant salt will often have a pH that is also neutral.

Neutralisation has a variety of practical uses. For example, as most plants grow best at neutral pH7, acidic or alkaline soil can be treated with chemicals to change its pH. In the case of acidic soil this is often calcium carbonate (chalk) or calcium oxide (quicklime). Another example is the human stomach, which contains hydrochloric acid. However, too much can lead to indigestion, so the acid can be neutralised with a base such as an indigestion tablet.

How do an acid and base react to produce salt, water and heat?

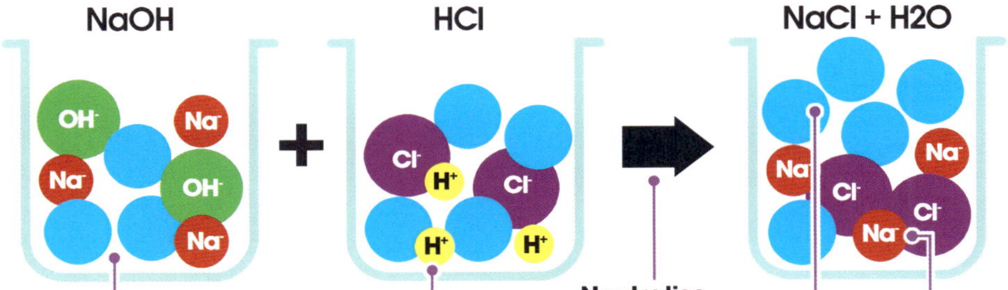

Sodium hydroxide
This strong alkali (a base soluble in water) has a pH of 13 or 14.

Hydrochloric acid
With a pH of 1 or 2, the H+ ions of this strong acid are removed by the alkali.

Neutralise
By removing the H+ ions, the alkali neutralises the acid and turns the ions into water.

Water
Neutral water of pH7 is produced.

NaCl
Neutral sodium chloride, or table salt, is also produced.

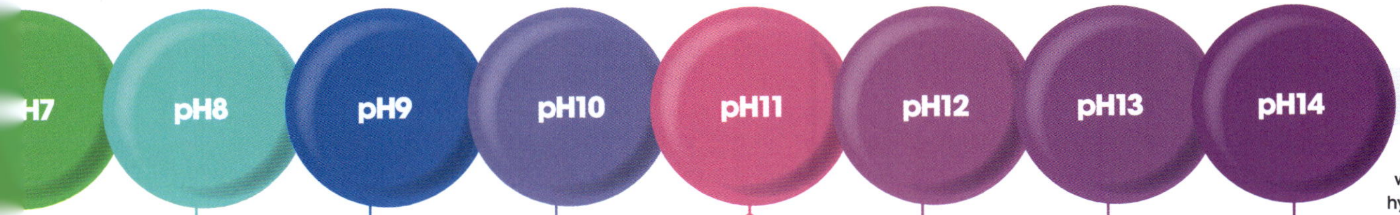

Base

A compound which 'accepts' hydrogen ions in an aqueous solution. Has ions of the opposite charge, eg, hydroxide (OH-), which is naturally found in water and negatively charged.

ed water
water is tral as it s the same of positive s negative ough most sn't pure in sense.

Baking soda
A slightly salty substance that is used as a base in foods to regulate the pH if something is too acidic.

Toothpaste
Acidic toothpaste can put enamel at risk of decay, so a weak base, such as sodium hydroxide, is added in order to regulate the pH.

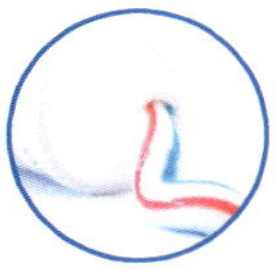

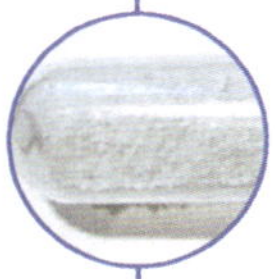

Milk of magnesia
A weak base of magnesium hydroxide in water, used to ease stomach aches caused by too much acid.

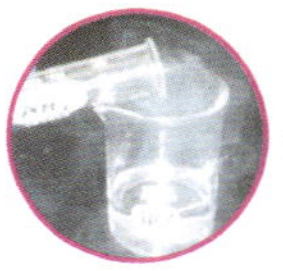

Ammonia solution
When placed in water, ammonia removes protons from a small fraction of the water to form ammonium and hydroxide. It is used in many cleaning products for its basic properties.

Bleach
Can contain sodium hypochlorite at different strengths, making it a strong caustic base.

Caustic oven cleaner
Heavy-duty oven cleaners can be really caustic and corrosive, helping to break down fat and grease.

Caustic soda
Chemically known as sodium hydroxide, in its purest form it is a white solid and can cause severe burns due to its high alkalinity.

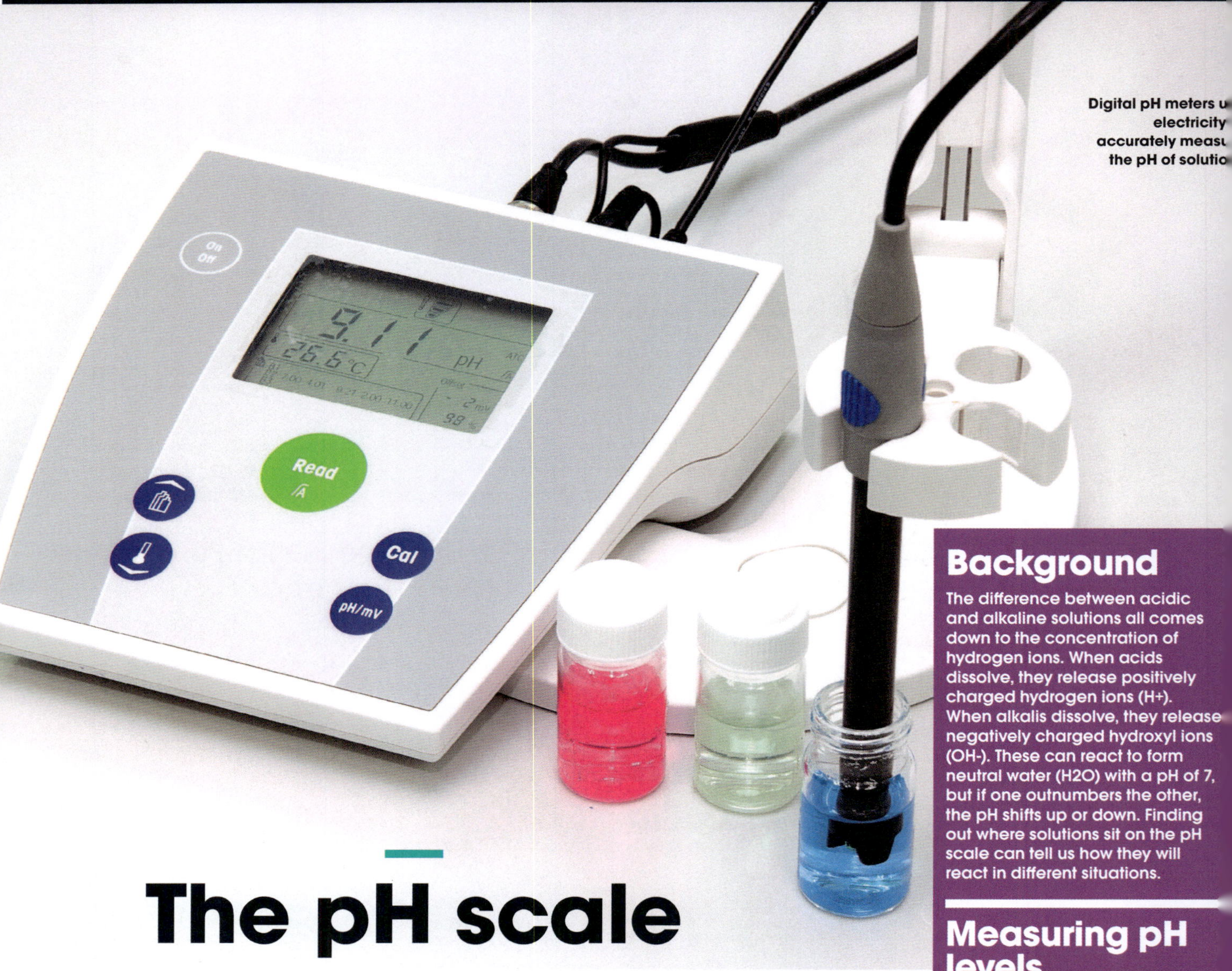

Digital pH meters u electricity accurately measu the pH of solutio

The pH scale

Sort your acids from your alkalis with this simple 15-point chart

The pH of a solution tells you what concentration of hydrogen ions it contains. The scale runs from 0 to 14. In mathematical terms the pH number is the negative logarithm of the concentration of hydrogen ions in a solution. Put simply, for every step up you go on the pH scale, there are ten times fewer hydrogen ions in solution. At pH0, the concentration of hydrogen ions is 1 mol dm-3, at pH1, it's 0.1 mol dm-3, at pH2, 0.01 mol dm-3 and so on.

Indicator solutions reveal the pH of unknown liquids because they change colour when they react with H+ or OH- ions. The most well known indicator is litmus, which turns red in acids and blue in alkaline solutions, but a universal indicator gives an even clearer picture. It contains several indicator solutions, creating a rainbow of colours to represent different pH values.

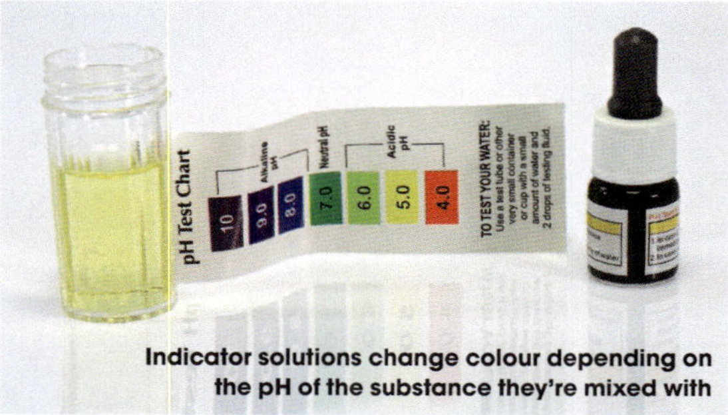

Indicator solutions change colour depending on the pH of the substance they're mixed with

Background

The difference between acidic and alkaline solutions all comes down to the concentration of hydrogen ions. When acids dissolve, they release positively charged hydrogen ions (H+). When alkalis dissolve, they release negatively charged hydroxyl ions (OH-). These can react to form neutral water (H2O) with a pH of 7, but if one outnumbers the other, the pH shifts up or down. Finding out where solutions sit on the pH scale can tell us how they will react in different situations.

Measuring pH levels

One of the most accurate ways to measure pH is with a digital pH meter. Rather than rely on matching the colours of an indicator solution, it gives a digita readout, measuring fractions of a point on the pH chart.

A digital pH meter has a probe that contains two electrodes; a glass electrode and a reference electrode. The glass electrode contains a solution of potassium chloride with a set number of hydrogen ions and a pH of 7. Whe a current passes through the prob hydrogen ions from the glass electrode and hydrogen ions fron the test solution interact with the glass. If the concentrations of hydrogen ions differ, it creates a difference in voltage. The meter measures this 'potential differenc to calculate the concentration of hydrogen ions in the test solution.

Everyday pH

Universal indicator can tell you where everyday solutions sit on the pH scale

Acid indicator

niversal indicator
rns from yellow to
orange to red as
solutions become
more acidic.

Neutral indicator

Universal indicator turns green when solutions are neutral – pH7.

Alkaline indicator

niversal indicator
urns from blue to
avy to purple as
lutions become
more alkaline.

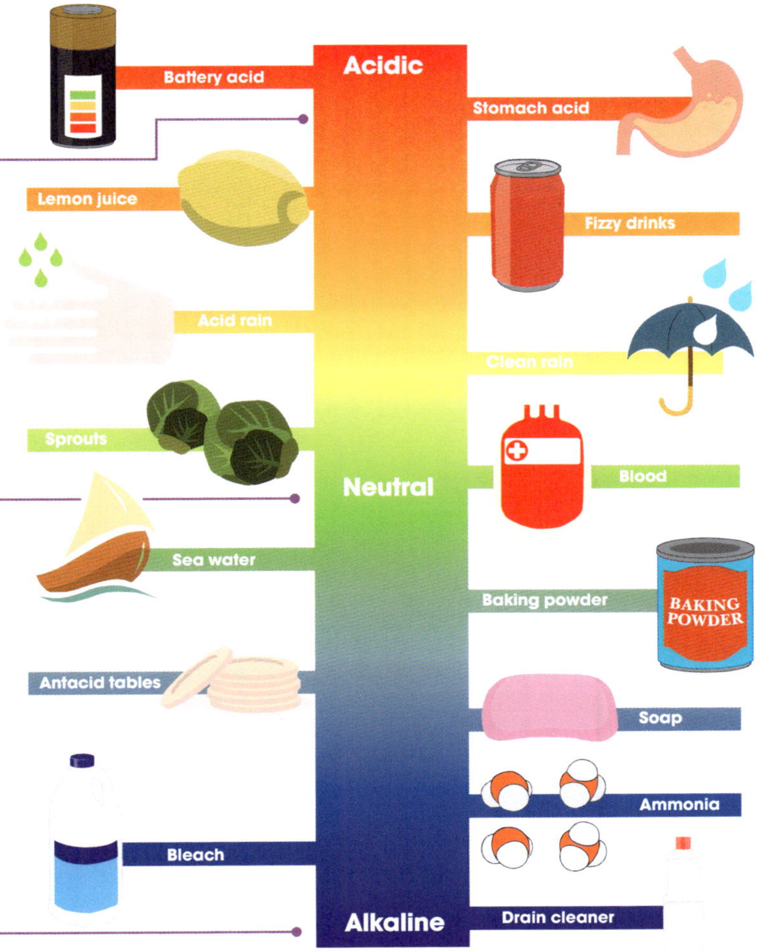

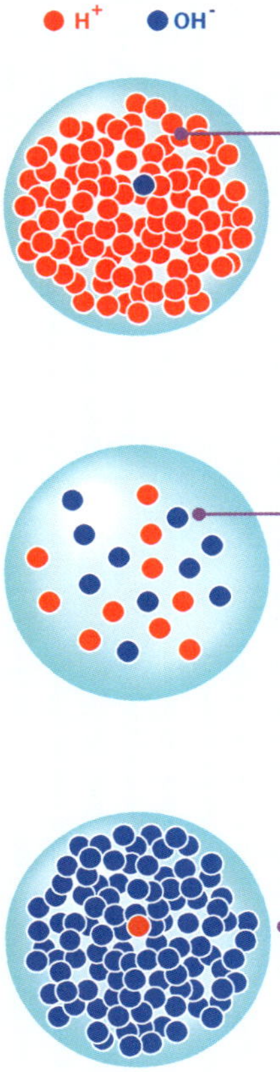

Acidic

Solutions with a pH value below 7 are acidic. The lower the value, the more free hydrogen ions there are.

Neutral

Pure water sits at the centre of the pH scale, with equal numbers of hydrogen and hydroxyl ions.

Alkaline

Solutions with a pH value above 7 are alkaline. The higher the value, the fewer free hydrogen ions there are.

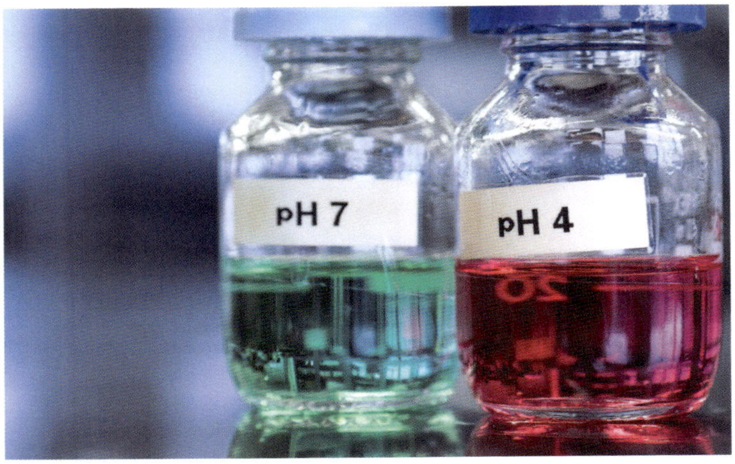

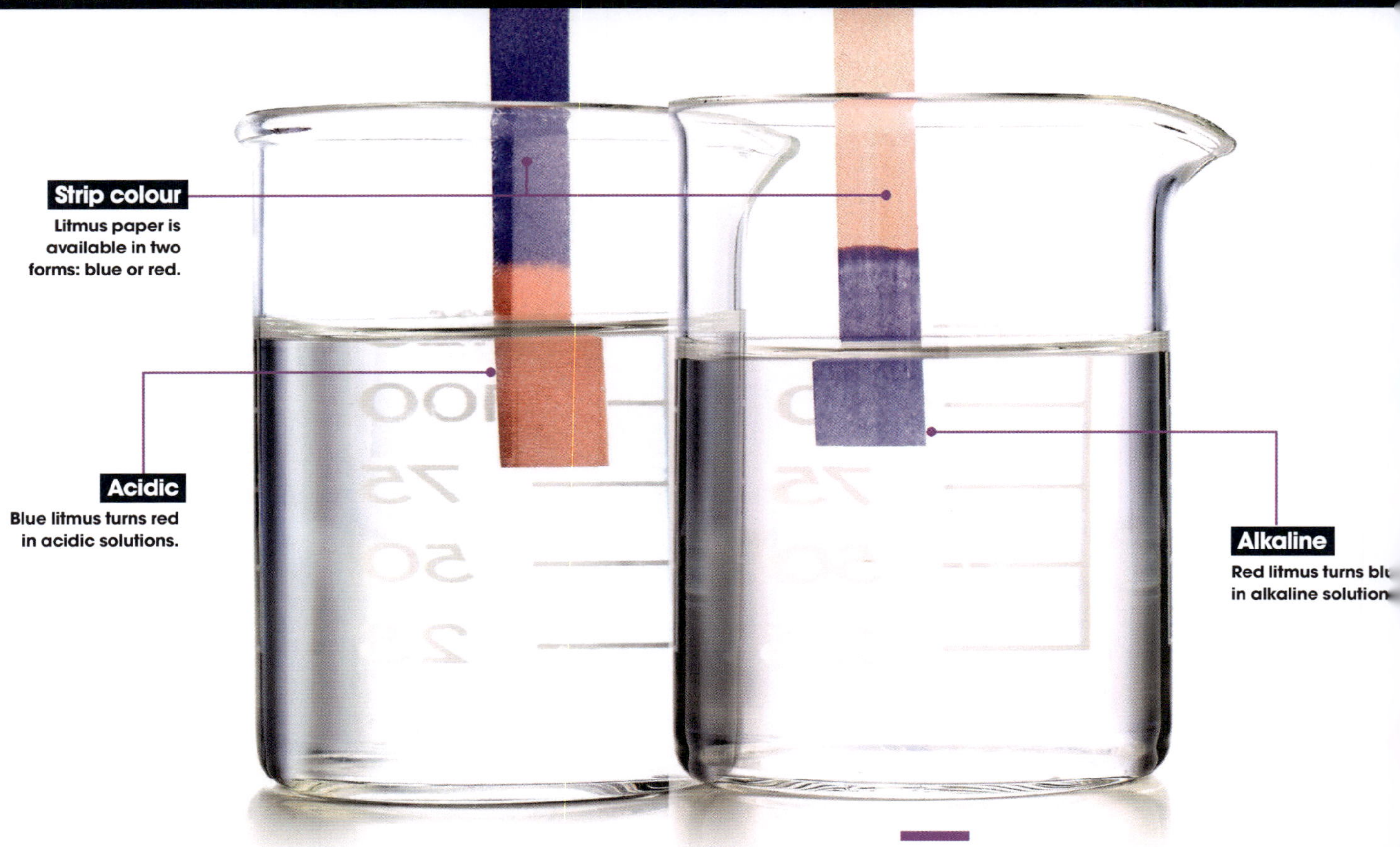

How litmus paper reveals pH

The science behind the simple tool able to identify a solution's acidity

In this image, blue litmus paper turns red in the presence of citric acid found in citrus fruits

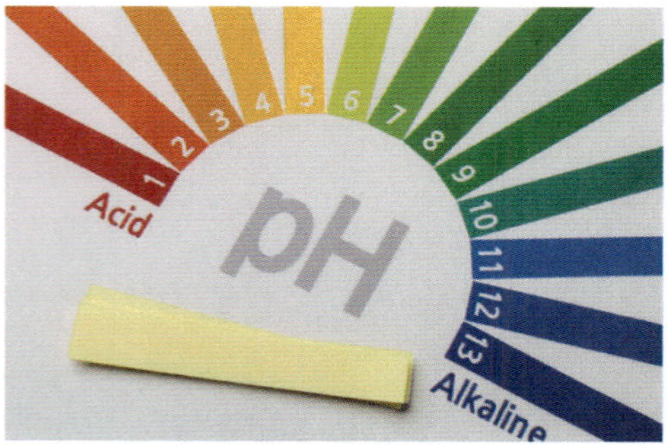

Not to be confused with litmus, universal indicator can also detect pH, but is able to show exactly how acidic, alkaline or neutral a solution is

Most young chemists will know litmus paper from science lessons, and older chemists for assessing their home brew. This colour-changing tool is used to find the acidity or alkalinity of a solution – acidic solutions turn the paper red, and alkaline (basic) solutions turn the paper blue. This provides a visual indication of the pH, which is a measure of the concentration of hydrogen ions in a solution.

Water molecules can be broken down into positively charged hydrogen ions and negatively charged hydroxide ions. Acidic solutions have higher concentrations of hydrogen ions than hydroxide ions, while the opposite is true of basic solutions. In water and other neutral solutions, the concentration of hydrogen and hydroxide ions is equal.

The colour-changing properties of the paper are due to crushed and fermented lichen that has been dried onto the surface. Lichens are a diverse group of organisms, many of which possess large, light-absorbing molecule called chromophores. These absorb differing wavelengths of light depending on their atom structure, which can be altered by the presen of the ions found in acidic and basic solutions.

Wavelengths at the blue end of the visible spectrum are absorbed when the chromoph react with hydrogen atoms, and wavelengths the red end of the spectrum are absorbed wh the chromophores react with hydroxide ions present in bases.

Depending on which wavelengths of light are absorbed, the paper will appear a different colour. We can then use this colour determine if the solution is acidic or basic.

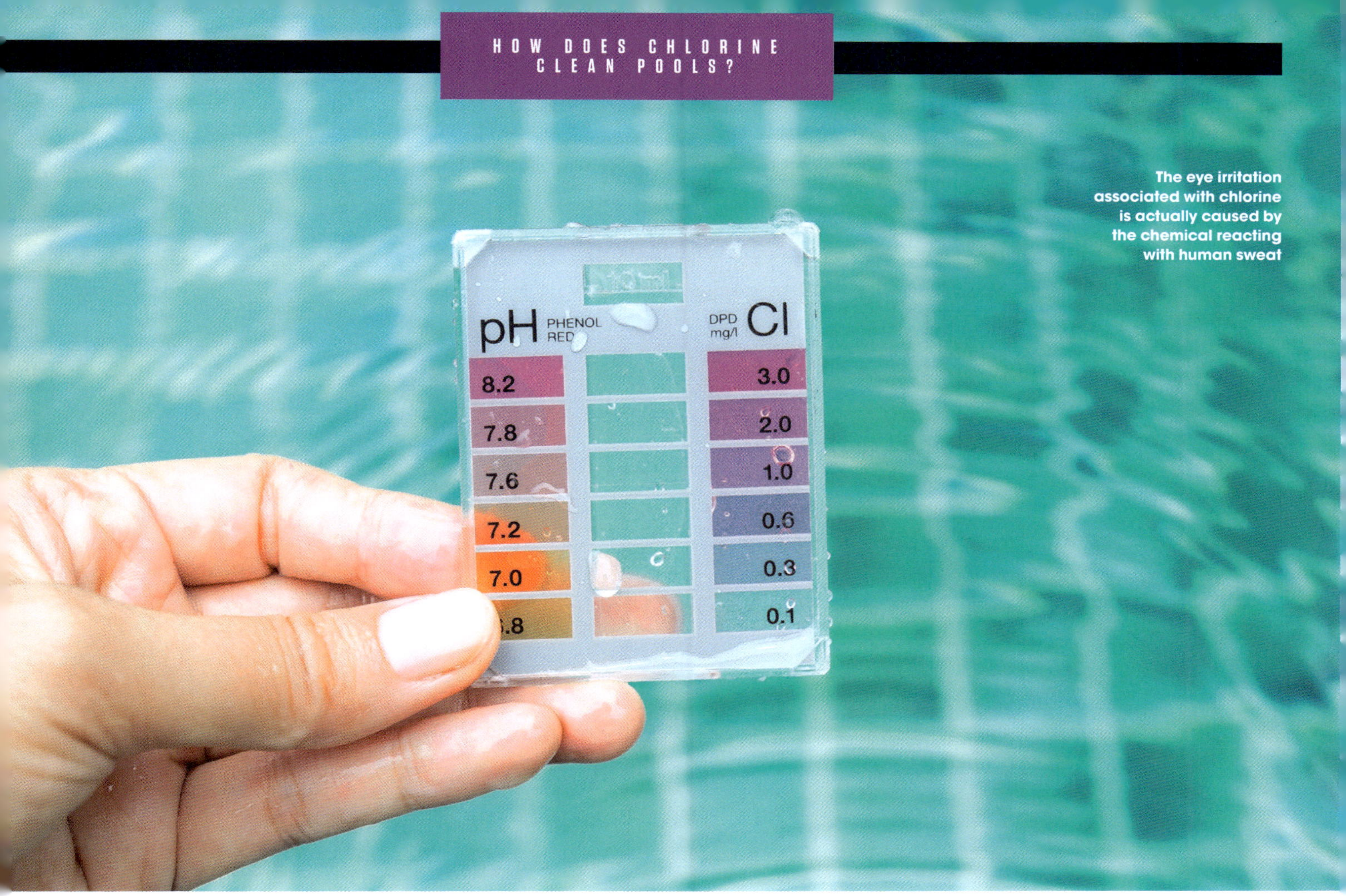

The eye irritation associated with chlorine is actually caused by the chemical reacting with human sweat

How does chlorine clean swimming pools?

The chemical reaction that stops swimmers from getting sick

With hundreds of people taking a dip each day, public swimming pools are a prime place for water-borne diseases, such as E. coli, to spread. Thankfully, a clever chemical can be used to keep your splashing sanitary.

When chlorine is added to pool water, it dissolves to form hypochlorous acid, a weak acid that acts as a disinfectant. It kills bacteria through a process called oxidation, which destroys protein structures inside the bacterial cells, causing them to die. As the water becomes more acidic, the hypochlorous acid itself splits up to form hypochlorite ions, which are less efficient at cleaning water. Therefore, for more effective disinfection, as well as comfortable swimming, the acidity of the pool water must be continuously monitored and kept at a neutral pH between 7.2 and 7.8.

Although it is effective at cleaning pools, chlorinated water also has some irritating side effects for swimmers. However, it's not the chlorine itself that's directly responsible for the red eyes we sometimes experience after a visit to the pool. This is the result of ammonia compounds, found in human sweat and urine, reacting with the hypochlorous acid to form chloromines, a compound that's also responsible for that strong chemical smell in swimming pools.

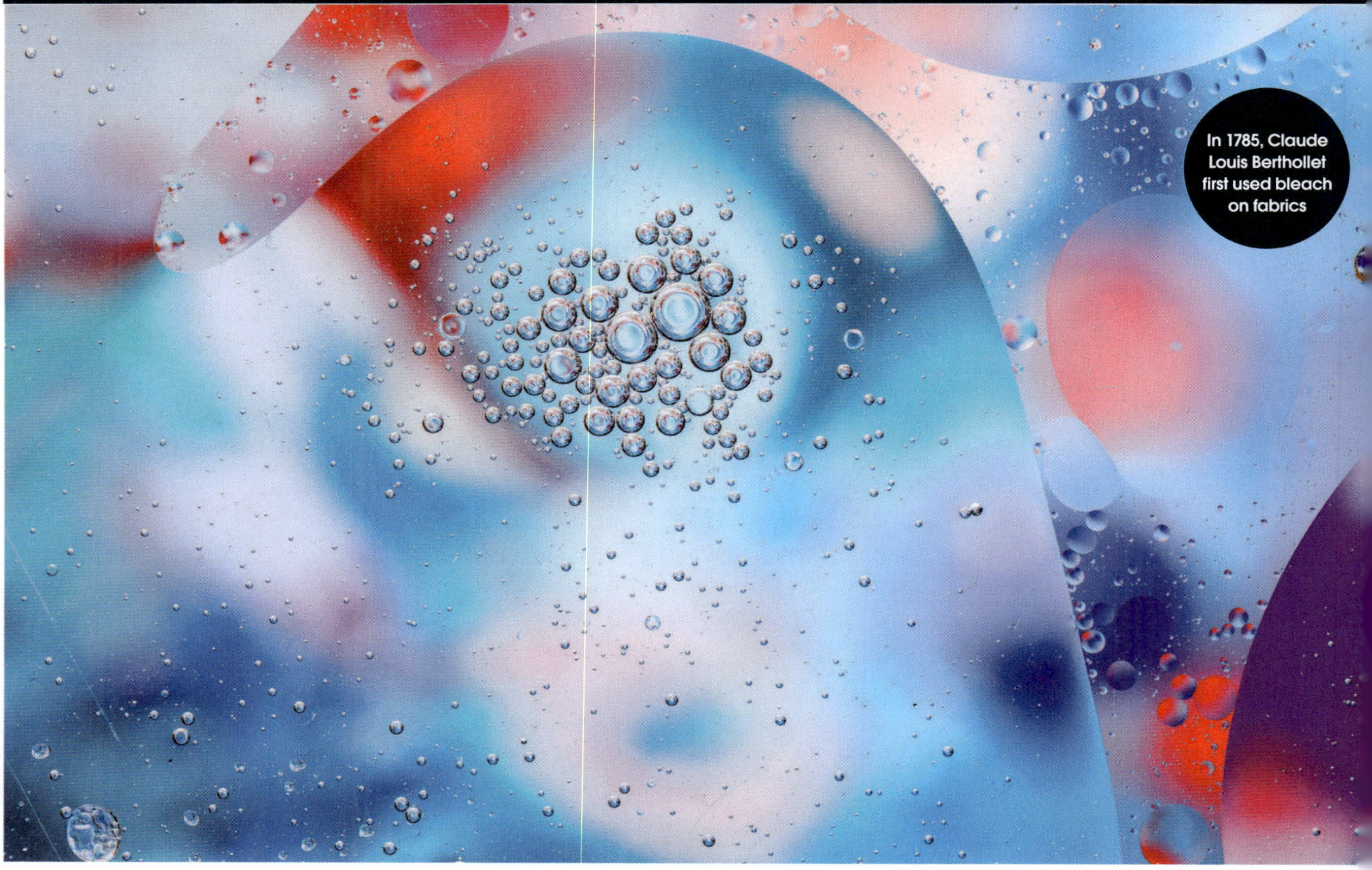

How bleach kills germ…

This cleaning cupboard staple tears through proteins, destroying bacteria and viruses in minutes

Household bleach is a bactericide; it kills bacteria on contact. It's one of the cheapest and most powerful microbe busters on the planet. But how does it work? The bleach under your sink is a solution of sodium hypochlorite, which has the chemical formula NaClO. That means that each molecule contains one sodium atom, one chlorine atom and one oxygen atom. In water it produces hypochlorous acid (HOCl), which can steal electrons from other molecules. It uses this ability to attack bacteria and viruses, breaking their proteins apart and splitting them open.

Hypochlorous acid also damages pigment molecules, turning them white, which is where bleach gets its name from. In its pure form, bleach is extremely corrosive. It has a pH of 13, making it just about as alkaline as you can get. Luckily, the bottle under your sink also contains water, lowering the pH to a slightly safer 11.

It doesn't take much bleach to completely sterilise a surface. Your bottle at home is only five per cent bleach, and you need to dilute it even further before you can use it safely. You only need 15 millilitres in 3.8 litres of water to sterilise a table that has been in contact with dirty water. To kill mould, you need a bit more – 240 millilitres in 3.8 litres. Bleach is so reactive that you have to use it within 24 hours of mixing it up or it won't be nearly as effective. It breaks down rapidly into salt and water, especially if it's warm, making it a much less powerful microbe killer. To keep it working a… its best, store it in a cool, dark cupboard and only dilute just as much as you need to get th… job done.

It takes between 10 and 60 minutes for bleach to kill germs

ow is bleach made?

Bleach is an industrial chemical, and it's produced on an industrial scale. It involves two methods: the chlor-alkali process and the Hooker process. The chlor-alkali ocess produces the raw materials for bleach oduction: sodium hydroxide, also known as ustic soda or lye, and chlorine gas. It begins th salty water, also known as brine. The lor-alkali process uses a technique called electrolysis to break the chemicals in brine apart and remake them into something new. When a current passes through the liquid, it turns water (H2O) and salt (NaCl) into sodium hydroxide (NaOH), chlorine (Cl2) and hydrogen (H2). Two of these three ingredients go on to the next stage, the Hooker process. In the Hooker process, the sodium hydroxide is cooled and the chlorine gas is bubbled through. This produces NaClO, also known as sodium hypochlorite, the active ingredient in household bleach.

Bleach is made from chlorine gas and sodium hydroxide

ow bleach kills

ng bleach is a great way to destroy bacteria

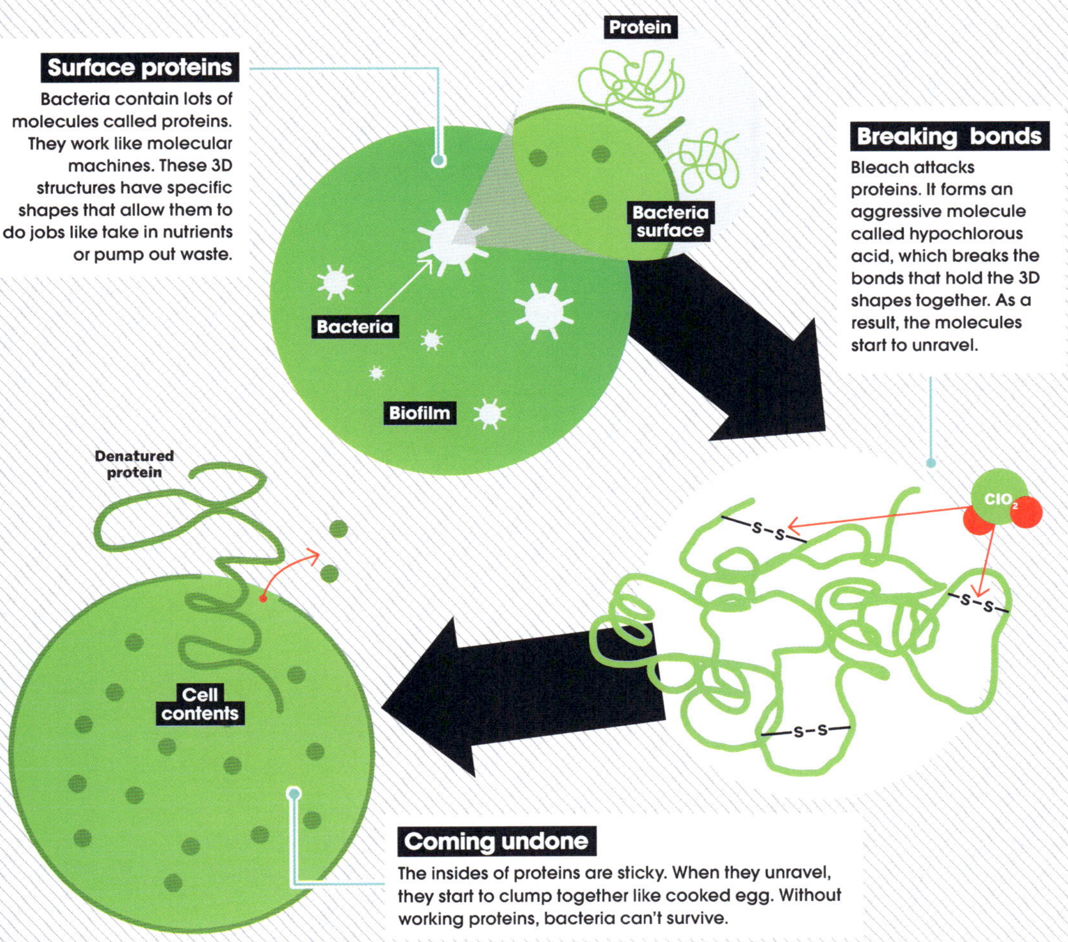

Surface proteins

Bacteria contain lots of molecules called proteins. They work like molecular machines. These 3D structures have specific shapes that allow them to do jobs like take in nutrients or pump out waste.

Breaking bonds

Bleach attacks proteins. It forms an aggressive molecule called hypochlorous acid, which breaks the bonds that hold the 3D shapes together. As a result, the molecules start to unravel.

Coming undone

The insides of proteins are sticky. When they unravel, they start to clump together like cooked egg. Without working proteins, bacteria can't survive.

> "It doesn't take much bleach to completely sterilise a surface"

5 Myths about bleach busted

1 More is better
Bleach is a powerful and dangerous chemical, so more is definitely not better. It only takes one tablespoonful diluted in a gallon of water to kill germs.

2 It causes cancer
While bleach can damage cells and can harm the lungs after severe or long-term exposure, there's no evidence that it can cause cancer in humans.

3 It contains chlorine gas
Bleach contains the chemical sodium hypochlorite, which contains the element chlorine. But the bottle doesn't contain any free chlorine gas.

4 It creates dioxins
Dioxins are toxic ring-shaped chemicals that cause health problems in humans. They're made during the industrial bleaching of paper, but they're not made by household bleach.

5 Drinking it kills coronavirus
Bleach can kill the SARS-CoV-2 coronavirus on surfaces, but it doesn't work on or inside the body. It causes serious harm to human cells.

Fluoride explained

How this mineral can protect your pearly whites

Fluoride is a mineral found naturally in water, soil, and in trace amounts in tea and fish. Fluoride can be used as a biochemical agent, but its main use is to help prevent the formation of tooth decay. Around 70 years ago, researchers found that people who drank water containing more fluoride had less tooth decay. This discovery led to the introduction of schemes around the world to add fluoride to water supplies where the level was low.

The substance also became commonplace in many brands of toothpaste. Fluoride protects teeth by encouraging a stronger enamel to form that's more resistant to acid attack. It also reduces plaque bacteria's ability to produce acid – the primary cause of tooth decay. Unfortunately, it seems that you can have too much of a good thing. In the US, the fluoride in water was found to cause white splotches on children's teeth – a condition known as dental fluorosis – so the government lowered fluoride levels. However, as long as the standard guidelines are followed, fluoride really can be beneficial to your dental health.

Fluoride is used in toothpaste to help fight decay

Making fertiliser

How the Haber process helps us grow food

The Haber process is an efficient way of producing ammonia for use in fertilisers and household products. Ammonia has helped to sustain food production for billions of people, but its use in explosives has reportedly resulted in the death of 150 million. It's for this reason that some scientists say ammonia changed the course of the 20th century more than electricity or television.

Industrial production of ammonia

The ins and outs of the Haber process

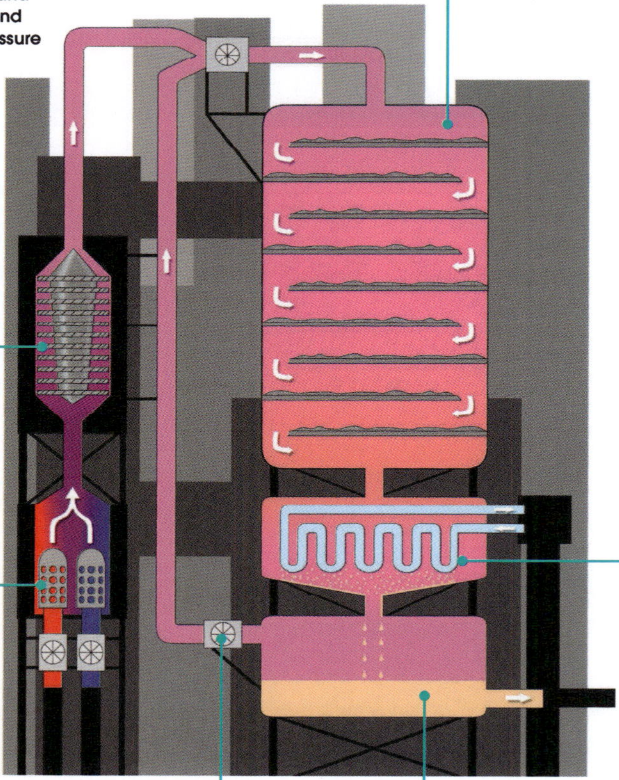

Gas compression

The clean hydrogen and nitrogen are mixed and compressed at a pressure of 200 atmospheres.

Reaction tower

The mixture of nitrogen and hydrogen passes through the reaction tower, where iron and hig[h] temperatures of 450°C (842°F) catalyse the formation of ammon[ia]

Cooling loo[p]

The ammonia produced in the reaction tower i[s] gaseous. The cooling loop w[orks] to condense it i[nto] a liquid.

Nitrogen and hydrogen cleaning

Before the process can begin, the nitrogen and hydrogen need to be cleaned and purified.

Gas recycling

The process doesn't react all the hydrogen and nitrogen together; therefore any leftover gas can be recycled back into the reaction tower.

Ammonia collection

Once condensed, the liqui[d] ammonia is piped off for collection, and can be store[d] in a refrigerator.

Where does acid rain come from?

We've all seen the effects of acid rain on limestone statues, but how does this damaging substance form?

All rainwater is a little bit acidic, because the carbon dioxide present in the atmosphere
solves in water and forms
rbonic acid. Stronger acid
n, however, can damage stone
uctures and can also be harmful
crops, as well as polluting
aterways. It forms in the
mosphere when
isonous gases
itted by human
ivities combine
th the moisture
thin rain clouds.
Fossil-fuelled power stations
d petrol/diesel vehicles give
chemical pollutants – mainly
phur dioxide (SO2) and nitrogen
des (NOx) – which when mixed
h the water in the air react and
n acidic.

Acid rain in action

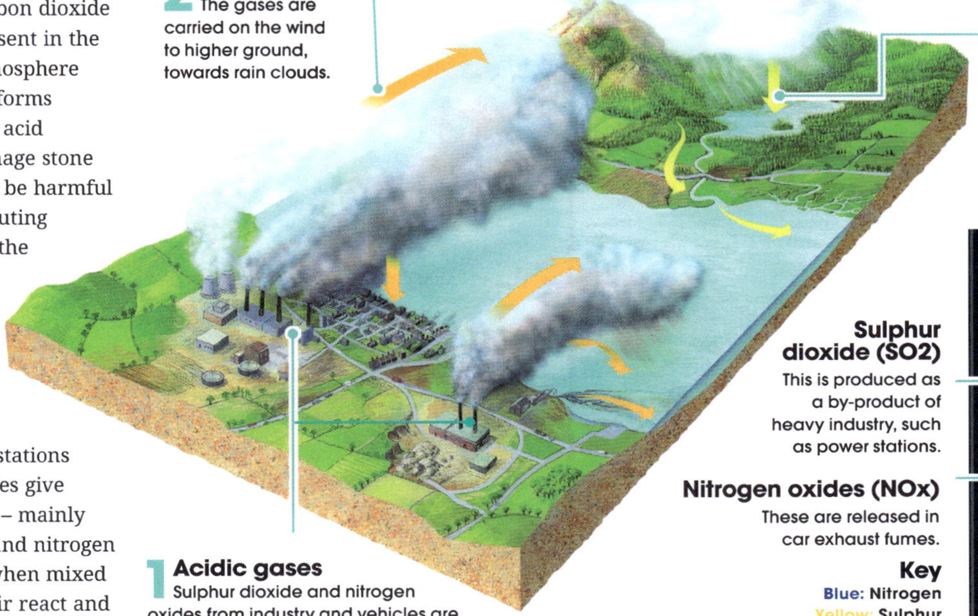

Oxidation of sulphur and nitrogen

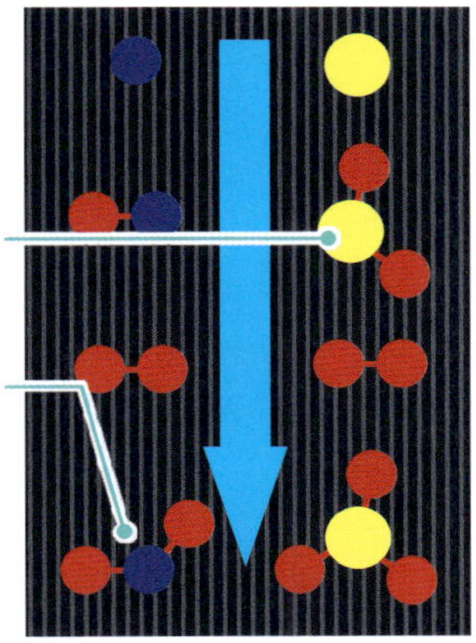

The smell of rain

nd out why precipitation creates a distinctive roma that's the same all over the world

It's possible to smell rain before it has even fallen. Lightning has the power to split atmospheric nitrogen and oxygen molecules into individual atoms. These atoms react to form nitric oxide, which in turn can interact with other chemicals to form ozone – the aroma of which is a bit like chlorine and a specific smell we associate with rain. When the scent
rries on the wind, we can predict the rain before it falls.
nother smell associated with rain is petrichor – a term coined by
ouple of Australian scientists in the mid-Sixties. After a dry spell of
ather, the first rain that falls brings with it a very particular aroma that
ne same no matter where you are. Two chemicals are responsible for
production of this indescribable odour called petrichor. One of the two
micals is released by a specific bacteria found in the earth; the other is
oil secreted by thirsty plants. These compounds combine on the ground
, when it rains, the smell of petrichor will fill your nostrils.

Catalysts can aid the production of oxygen by decomposing hydrogen peroxide

Catalysts

What's really at the middle of a chain reaction?

Reactions can be as slow as the acid in your stomach digesting food over a period of several hours, or as quick and dramatic as a piece of magnesium setting alight. However, there are ways you can alter and – to an extent – control the speed of a reaction and that is by using a catalyst.

Much like stepping on the accelerator pedal of a car makes it go faster, using a catalyst has the ability to speed up a chemical reaction by providing a surface ideal for a reaction to occur. Particles are attracted to the surface of the catalyst, allowing many more successful collisions and therefore reactions. Substances that slow down a reaction work the opposite way, reducing the success rate of collisions, and are called inhibitors.

A catalyst will generally be made of a transition metal like iron or nickel because they can comfortably give and receive electrons, a function that is crucial to many chemical reactions.

What is dry ice?

How do rock stars and actors get their atmospheric assistance?

If you've ever been to any kind of stage show, there's a decent chance you'll have seen dry ice being used. This is the stuff that's pumped onto the stage to simulate mist and fog, or generally make the place just look awesome, but what is it and how is it made?

Dry ice is actually highly pressurised liquid-carbon dioxide that has been frozen to a temperature of -78.5 degrees Celsius (-109.3 degrees Fahrenheit). The main benefit of using carbon dioxide is that when it warms up it sublimates, which means it changes from sol to gas without becoming a liquid. So if you warm it up and pump it across a stage, it wil turn instantly to cool-looking mist, rather tha just soaking the floor.

It is also really useful for packing frozen goods that need to stay cool because when it warms up, it turns gaseous rather than wet.

Making dry ice

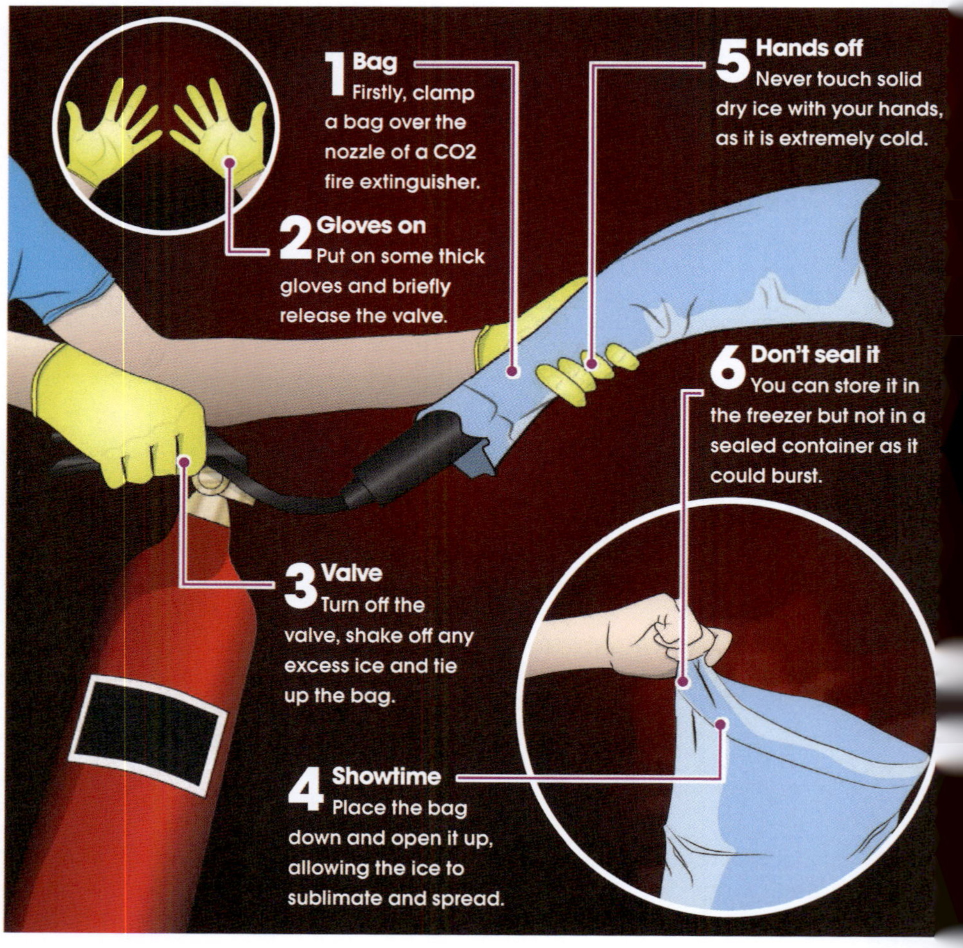

Mercury

What makes this metal element so unique, and why do we normally see it as a liquid?

Also known as quicksilver, mercury is a heavy metal occurring naturally on Earth. It is the most common of five metals that are a liquid at room temperature. Compared
ɔther metals, mercury is a poor conductor of
ıt but a good carrier of electricity. It is most
ımonly found in rocks such as mercuric
ɔhide inside the ore cinnabar. Mercury is
racted by heating and compressing this ore.
nside a mercury atom, the electrons are very
ıtly held together. For this reason the bonds
ween the atoms are very weak as they do not
re electrons easily. The small amount of heat
sent at room temperature is enough to break
bonds between the atoms, which is why
rcury is a liquid at such a low temperature.
er metals, such as gold, have much stronger
ıds between their atoms and therefore require a
more energy (or heat) to become a liquid.

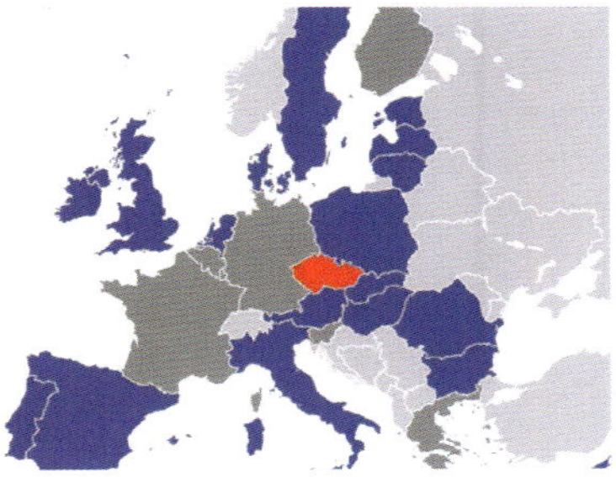

Mercury bans

Mercury thermometers can be dangerous, but where have they been banned?

- Banned
- Considering ban
- No ban
- Not in discussion

ere is mercury ıed?

- maden, Ciudad al, Spain
- ija, former goslavia
- nta Clara Co, lifornia
- ina
- rgyzstan
- scany, Italy
- geria

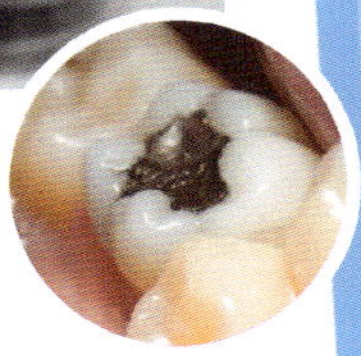

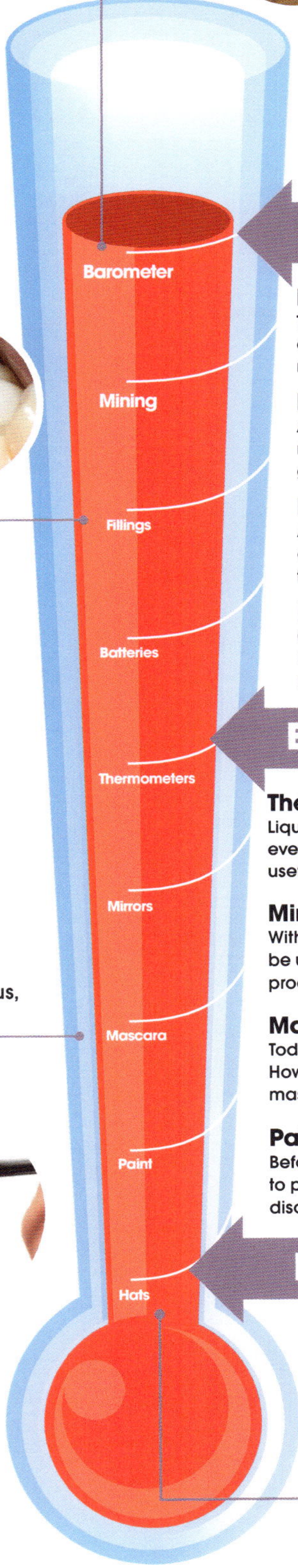

What we use mercury for today...

There have been calls to ban mercury, as it can be poisonous as a vapour. While it's slowly being phased out, it is still in use

Still in use

Barometer
The high density and surface tension of liquid mercury is perfect for measuring atmospheric pressure.

Mining
As mercury combines with other metals, it can be used to extract gold and silver from a mine.

Fillings
A mixture of liquid mercury and other materials can be used to fill in the holes in our teeth.

Batteries
Once used to prevent zinc corroding in regular batteries, liquid mercury is now only used in button batteries.

Being replaced

Thermometers
Liquid mercury expands and contracts evenly with a change in temperature, a useful feature for thermometers.

Mirrors
With excellent reflectively, mercury can be used as a compound with other products to make mirrors.

Mascara
Today mercury is rarely used in make-up. However, it was once an ingredient in mascara for its binding properties.

Paint
Before 1992, mercury was used in paint to prevent fungus growth, until it was discovered it could be poisonous.

No longer used

Hats
Mercury was used in the felting process to make hats in the mid-19th century, coining the phrase "mad as a hatter".

Gold: the rich element

How this versatile, unique metal forms, and how it's extracted and used today

Gold represents a tiny fraction of the elements in the known universe. The reason for its rarity is owed to the incomprehensible amount of energy needed for its formation. Gold is formed within stars, but only in those that are exploding in giant supernovae or incredibly dense ones that have come together in monstrously powerful collisions.

Stars generate energy through the power of fusion, where smaller elements are fused, or combined together into heavier elements. To start with, a star may be mostly hydrogen, the lightest element. The process of fusion under the immense pressure and heat in the star's core will generate helium. When hydrogen runs low and the star begins to reach the next phase of its life cycle, it will fuse helium into the next heavier element, and so on. This process continues until the element iron is made, where the balance suddenly shifts – fusing iron doesn't create energy, it consumes it. With no means of generating internal energy to counteract its own immense pressure and gravity, the star begins to collapse onto itself. If the star is large enough, the result is a supernova – a massive stellar explosion. Heavier elements are formed from the incredible energy generated during this process, including gold.

Scientists believe that gold forms following a collision between neutron stars

Very few cosmic events can generate the levels of concentrated energy as within a collapsing aged star, but one that can surpass it is collisions between neutron stars. Neutron stars are small and incredibly dense stars that fell short of becoming black holes. They generate huge gravity and can enter another's orbit. This can result in collisions of enormous magnitudes, generating enough energy to form gold and other heavier metals.

Once flung out into the void of space, gold and other elements slowly came together to form the nexus of our fledgling planet. Over time, as the Earth cooled and separated into layers of core, mantle, and crust, the gold trapped deep within was forced towards the surface. Several theories have been suggested as to how this process occurs, but a shared theme is that heat and pressure forced liquid water up, carrying dissolved gold as it travelled. As the water cooled, the gold precipitated out of solution, forming veins or lodes of concentrated gold.

Over time, some gold travelled through the conduit of water yet further, up to the surface and into rivers. Here small nuggets of gold ran downstream. Perhaps the very first piece was encountered by a fisherman who spied a glint, dipping his hand into the water to recover the chunk of shiny metallic yellow. Thousands of years after this first occurred, gold prospectors and miners would treat gold nuggets in water as a sign of a lode nearby, and they'd busily get to penetrating the ground in search of gold veins to mine. Throughout the history of humanity, many have died for gold, and many have killed for it. But despite that, the story of where gold has taken our species is perhaps not as fascinating as where the element began.

Did you know?
Gold's purity can be measured in karats – 24k gold is entirely pure, but 14k is more common in jewellery

Unexpected find

Gold plays a strong role in Australian history. In the late-19th century, so many flocked to the country to take part in its booming 'gold rush' that the population tripled. Owing to pervasive deposits, it's still mined today. However, one company, Evolution Mining, found a different treasure in its hunt for gold. When drilling into the Australian Outback's surface in search of gold deposits, miners instead unearthed sheets of stone that resembled 'shatter cones', which form on the outer rims of impact craters. They followed up this finding with advanced mapping techniques, confirming the uncovering of a three-mile-wide meteorite crater.

Australia is home to multiple meteorite impact craters

Gold's natural purity and attractive colour have captivated human civilisations for thousands of years

4 common uses of gold

1 Medals and jewellery

Due to its beautiful colour, its rarity and its malleability, enabling it to be folded into all manner of shapes, gold has been a coveted jewellery material for millennia.

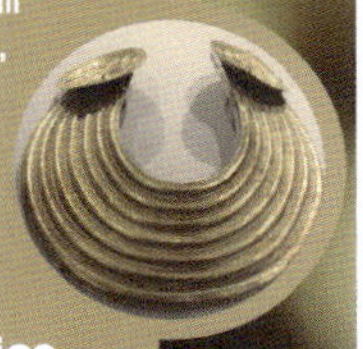

2 Electronics

Gold doesn't rust and it's a good conductor of electricity, making it a useful item for modern-day electronics. Gold coatings are applied to connectors, both for terrestrial needs and for use in spacecraft.

3 Dentistry

As well as being a hard-wearing metal that is pliable and doesn't rust, gold is also non-toxic. This makes it an excellent component of fillings used to fill cavities in dentistry.

4 Bullion

A rare metal that is non-toxic and remains untarnished by rust makes for an ideal medium of wealth exchange. Ancient civilisations used gold within their physical currency as coins, but today wealth is primarily invested into gold by purchasing bullion.

Aluminium

It's the most abundant metal in Earth's crust, yet it entirely escaped our notice until 1825

You might say that it was hidden in plain sight. Aluminium is a highly reactive metal, meaning it readily undergoes chemical reactions with other elements and compounds to form different substances. As a result, nearly all of the naturally occurring aluminium atoms on Earth ended up tucked away in the molecules of more than 270 different minerals, including gemstones like emeralds and rubies. While it's actually 8.2 per cent of the Earth's crust, making it the most common metal and third-most common element, behind oxygen and silicon, you would never know it's there without investigating on a chemical level.

The search was on in the mid-1700s, when chemists began experimenting with alum, a class of abundant chemical compounds. Alum compounds, such as potassium aluminium sulphate, were well known, going back at least to the ancient Greeks and Romans, who used them as an astringent to close wounds and a mordant to bind dye to cloth. Early chemical investigation of alum suggested that the compound included an unknown metal.

The trouble was that 18th-century chemists had no way to separate the mystery element from the rest of the atoms in the compound. In 1825 the Danish chemist Hans Christian Ørsted finally devised a chemical reaction th could extract it, but his process could only yield minuscule amounts at a time, making thorough experimentation difficult. Followin up on Ørsted's discovery, the German chemis Friedrich Wöhler developed a more effective process, and by 1845 he had produced enoug aluminium to demonstrate its basic properti However, Wöhler's method of extraction wa still far too troublesome and slow to support wide-scale production.

In 1856 the French chemist Henri Étienne Sainte-Claire Deville refined the process furt reducing the price of aluminium from $1,20

Aluminium extraction step-by-step

Commercial aluminium comes from bauxite ore, a mixture of hydrated aluminium oxide and hydrated iron oxide found in wide, flat layers near the Earth's surface

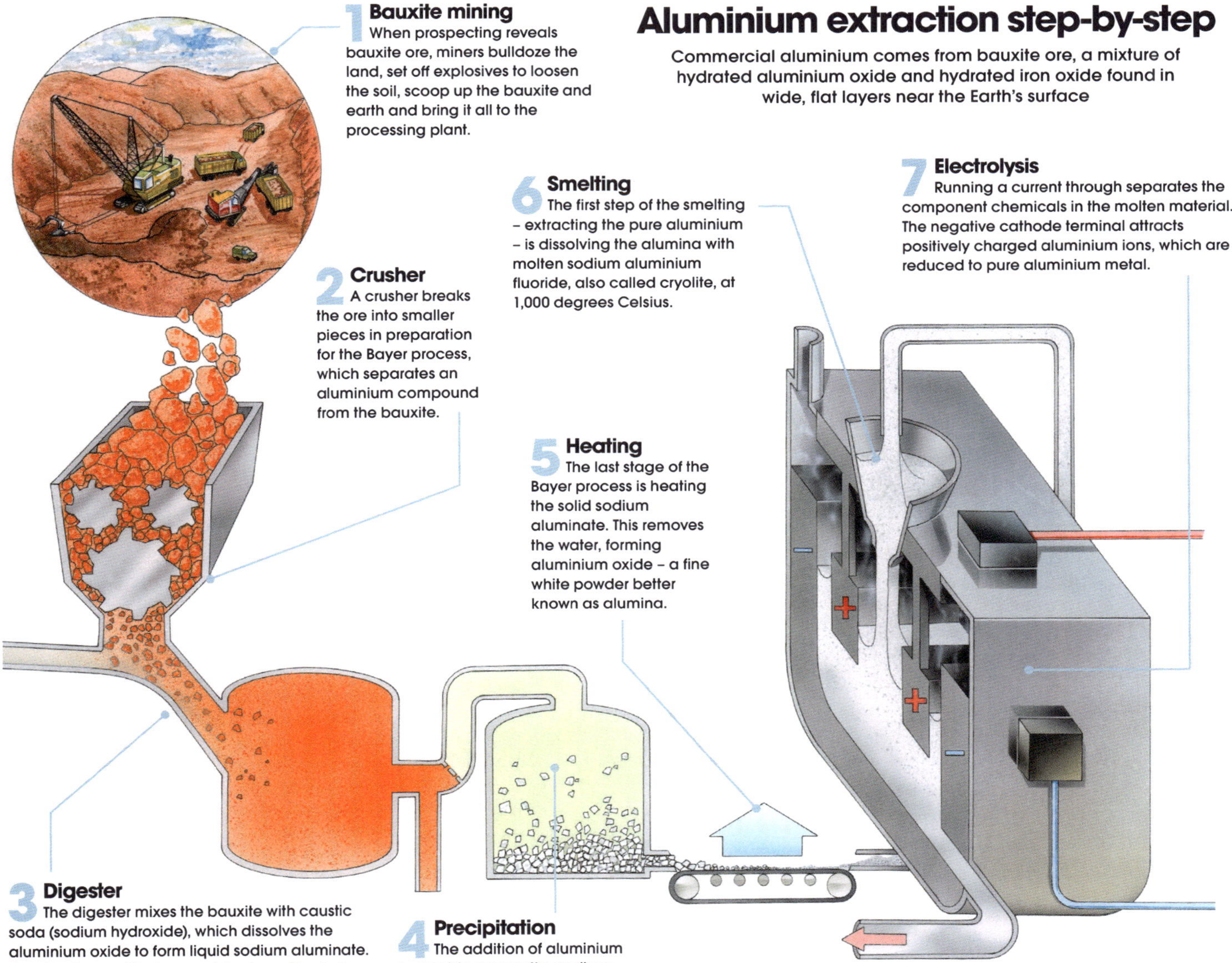

1 Bauxite mining
When prospecting reveals bauxite ore, miners bulldoze the land, set off explosives to loosen the soil, scoop up the bauxite and earth and bring it all to the processing plant.

2 Crusher
A crusher breaks the ore into smaller pieces in preparation for the Bayer process, which separates an aluminium compound from the bauxite.

3 Digester
The digester mixes the bauxite with caustic soda (sodium hydroxide), which dissolves the aluminium oxide to form liquid sodium aluminate. Clarification, or filtering, enables impurities to be removed from the solution.

4 Precipitation
The addition of aluminium hydroxide causes the sodium aluminate to precipitate into a solid.

5 Heating
The last stage of the Bayer process is heating the solid sodium aluminate. This removes the water, forming aluminium oxide – a fine white powder better known as alumina.

6 Smelting
The first step of the smelting – extracting the pure aluminium – is dissolving the alumina with molten sodium aluminium fluoride, also called cryolite, at 1,000 degrees Celsius.

7 Electrolysis
Running a current through separates the component chemicals in the molten material. The negative cathode terminal attracts positively charged aluminium ions, which are reduced to pure aluminium metal.

kilogram to $40, which was a huge drop, still very expensive. That all changed in the 0s, thanks to two key technological leaps. In 1886, American chemist Charles Martin l and French chemist Paul LT Héroult both ependently invented a process for extracting minium from aluminium oxide. The Hall-oult process relies on electrolysis, a means of aking down chemical compounds into their ponent elements using an electric current. basic idea is to conduct electricity from a itive terminal – an anode – to a negative minal – a cathode – via liquid or molten erial. Each terminal attracts and repels rged atoms (ions). The positively charged

Anatomy of aluminium

A look at the make-up of this metal on an atomic level

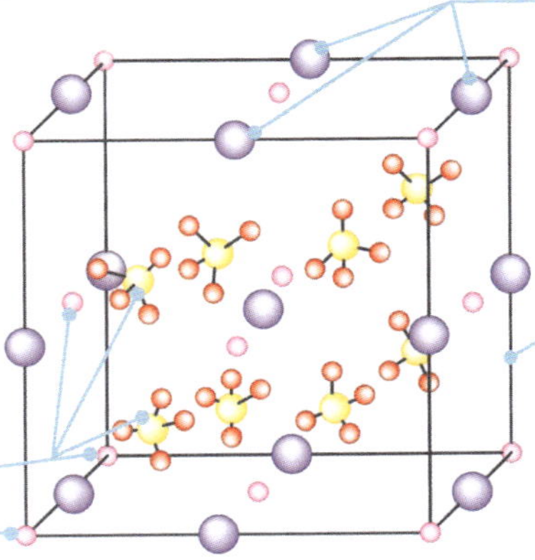

Aluminium atom
Each aluminium atom consists of a nucleus (yellow) orbited by 13 electrons, including three electrons in the outermost energy level. It is these valence electrons (red) that can be used for chemical bonding.

Chemical bonds
The individual aluminium atoms bond with one another by sharing valence electrons. In addition to the three it already has, each atom borrows extra electrons from other atoms.

Face-centred cube
The bonded atoms form a cubic structure. Together, many cubes form a crystal lattice that makes up solid aluminium material.

anode attracts negative ions and repels positive ions, and the cathode does the opposite.

Scientists had unsuccessfully tried producing aluminium through electrolysis since the 1800s. Hall and Héroult's breakthrough was dissolving aluminium oxide in molten cryolite – sodium aluminium fluoride. Applying an electric current to this material draws the positive aluminium ions to the cathode, typically the vat itself, made from iron lined with graphite.

Hot on their heels in 1888, Austrian chemist Karl Josef Bayer found a way to extract aluminium oxide from bauxite, a naturally occurring ore found in abundance in layers just below the Earth's surface. Geologists drill core samples in likely areas, and on locating bauxite they clear the ground above with bulldozers so that they can collect the ore beneath. Australia leads global bauxite mining, producing roughly one-third of the total ore.

Together the Hall-Héroult process and the Bayer process, both still in use, ushered in what could be called the 'Aluminium Age'. The metal's properties made it an instant hit. It's lightweight – about a third the weight of steel – but still strong. It's also very ductile, meaning it's easy to draw into a wire or flatten into a sheet, and it's malleable, making it relatively simple to bang it into just about any shape.

Add to that exceptional conduction of heat and electricity and you've got an incredibly versatile metal. But aluminium's greatest trick may be its resistance to corrosion. Like iron, aluminium is highly reactive to oxygen in the air, but the result of the oxidation reaction is very different. Oxygen and iron react to produce a flaky layer of rust, which falls away, revealing a lower layer of iron, which then oxidises to form yet more rust. In contrast, when aluminium encounters oxygen, the oxidation reaction produces an incredibly hard transparent oxide compound that essentially surrounds the aluminium with a shield that protects it from oxygen and other elements. And best of all, if this protective layer happens to get damaged, it will very quickly reform, reconstructing the shield.

> **"Scientists had unsuccessfully tried producing aluminium since the 1800s"**

Most aluminium products are actually made from an aluminium alloy – a combination of two metals. The combinations accentuate and amplify certain properties. For example, alloying aluminium with copper improves strength, while an alloy of aluminium and manganese improves resistance to corrosion.

You can turn aluminium into an infinite variety of products through a number of different manufacturing processes. You can cast it into any shape that you want by pouring it into a mould and then letting it cool. You can roll it into malleable sheets up to a minuscule 0.15 millimetres thick, commonly used in the kitchen. You can forge it to make it super-strong. You can machine it – cutting away material – to produce screws, bolts and other hardware. Finally, you can force it through a die to extrude it into a particular shape, including thin wire.

Aluminium also boasts another major superpower over many other metals: its recyclability. Recycling programs use old aluminium cans to make new ones at about 30 per cent the cost of making them from scratch. They shred old cans into pieces, melt them in a furnace, form rectangular blocks called ingots and then roll out the ingots into thin sheets from which new cans are cut. Believe it or not, this whole process can take just 60 days. Old car parts can undergo a similar process. Thanks to recycling, about 75 per cent of the aluminium ever produced is still in use today.

Aluminium consumption by market

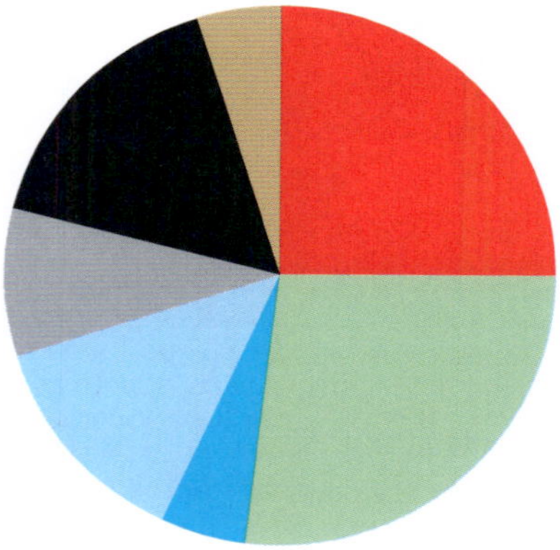

Building and construction	25%
Transport	27%
Consumer goods	5%
Electrical	13%
Machinery and equipment	9%
Foil and packaging	16%
Other	5%

Source: UC RUSAL (2014)

Rocket fue

While you might not be surprised to hear th NASA's Space Shuttles were made mainly fro aluminium, what you may not have realised is th they were also powered by aluminium inside th solid rocket boosters. When burned with oxyge atomised aluminium powder makes for a gre fuel. Aluminium powder accounts for about 16 p cent of solid rocket booster fue

World of aluminiun

It's durable, light and you can mould it into any shape you want. Little wonder it's everywhere...

ASM Space Lattice

Aluminium's high strength-to-weight ratio makes it an excellent dome material. Geodesic dome inventor Buckminster Fuller designed this 76-metre-diameter, 80-tonne aluminium structure for the American Society for Metals' (ASM) headquarters in Ohio.

Airstream trailers

The quintessential camping trailer took its design from 1920s aeroplane fuselages. Inventor Wally Byam opted for malleable aluminium, which he could shape into a fuel-efficient, aerodynamic form.

Ravensbourne University London

Aluminium's weather resistance and sculptural flexibility make it a popular material for building façades. Ravensbourne's building on London's Greenwich peninsula is covered in 28,000 aluminium tiles.

Top of the Washingt Monum

When the monument was approac completion in 1884, the lead engi selected the novel, relatively rare alumin for its 23-centimetre lightning rod pyra

Wider world

ISS

The US Destiny laboratory module is a major component of the International Space Station. The 8.5-metre pressurised unit is made from aluminium and represents the heart of the space station. Aluminium forms part of the outer debris shield too, which is tough enough to vaporise small particles of space junk.

Airbus A380

Aluminium has become the most important material in aerospace history. The world's largest commercial aircraft is 61 per cent aluminium alloy.

Morning coffee

Nespresso's airtight coffee capsules are made of aluminium to keep the product fresh, away from air, light and humidity.

Pots and pans

Much modern cookware includes aluminium, which boasts excellent thermal conductivity, but possible links to neurodegenerative diseases have made it somewhat controversial.

Everyday world

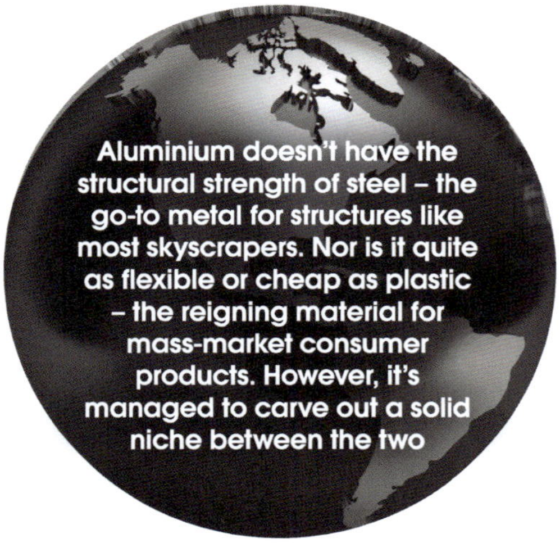

Aluminium doesn't have the structural strength of steel – the go-to metal for structures like most skyscrapers. Nor is it quite as flexible or cheap as plastic – the reigning material for mass-market consumer products. However, it's managed to carve out a solid niche between the two

Automobiles

Aluminium keeps this all-electric car lightweight, while still strong and rigid. Each car begins life as a 9,072-kilogram aluminium coil, which is then stamped into sections.

Phone lines

Aluminium is a great electrical conductor, like copper but much lighter. Its low weight makes it an ideal choice for use in elevated power and phone lines.

Computers

Many of Apple's devices are made of anodised aluminium, which not only polishes and toughens a product, but also provides a way of adding colour via the process of oxidation.

Kitchen foil

As a natural barrier to light, oxygen, moisture and just about anything airborne – including bacteria – flexible aluminium sheets are great food protectors.

Drinks cans

On top of being light and cheap, the king of aluminium products is 100 per cent recyclable. Around 113,200 cans are recycled every minute.

Burj Khalifa

The world's tallest human-made structure is also the highest installation whose architectural cladding consists of an aluminium and glazed façade. The total weight of the aluminium used is the same as five Airbus A380s, and the surface area of the curtain wall is 132,190 square metres.

Understanding uranium

Discover the chemistry of this naturally radioactive metal and how it is mined

Uranium is a metal of the actinide series: a group of large, radioactive elements, represented as an additional row on the periodic table. It is the heaviest of the naturally occurring elements and has 92 protons and 92 electrons. Like every element heavier than lead – which has 82 protons and 82 electrons – uranium is unstable and will gradually break down, emitting particles of ionising radiation.

Like all other elements, the number of neutrons in a uranium atom varies, and several isotopes of slightly different molecular weights exist in nature. All uranium isotopes are unstable, but only one is capable of establishing a nuclear chain reaction. The uranium isotope uranium-235 is a fissionable material; when it is struck by a neutron, the nucleus splits, releasing energy and more neutrons, sparking a chain reaction. It is this particular property that makes uranium an ideal fuel source for nuclear power plants, submarines and weapons, among other things.

However, the radioactive decay of uranium poses a significant danger to people and the environment. Not only is uranium radioactive, but it is also toxic and the products of its decay produce ionising radiation. Therefore, nuclear waste is handled with great care, first being stabilised and then encased in shielded vaults with reinforced, super-thick walls, either above or below ground.

Raising the radioactivity

Natural uranium only contains about 0.7 pe cent of the most useful isotope, uranium-23 which is insufficient to sustain a nuclear ch reaction. Through a process known as enrichment, however, this proportion can b increased up to around 90 per cent.

In order to enrich the uranium, the uranium-235 atoms must be separated fro the uranium-238 isotope that makes up ove 99 per cent of the natural metal. This is usu done using a gaseous compound of urani – uranium hexafluoride. The enrichment processes take advantage of the slight difference (less than two per cent) in mass between the two isotopes to gradually increase the amount of uranium-235. The depleted waste material is less radioactive than natural uranium, but is still hazardous and is kept in secure storage for safety.

"All uranium isotopes are unstable, but only one is capable of establishing a nuclear chain reaction"

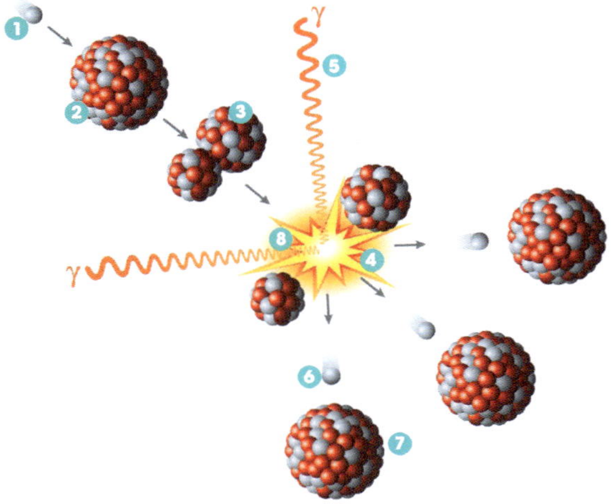

When uranium goes nuclear

Uranium-235 is unique among the element's isotopes as it can sustain a nuclear chain reaction, as shown here

1 Neutron
Subatomic particles with no electrical charge are introduced to a reactor containing uranium-235.

2 Collision
When a neutron smashes into a uranium nucleus it gains energy and becomes unstable.

3 Fission
The uranium nucleus breaks into two parts, forming two smaller elements and releasing spare neutrons.

4 Smaller elements
The fission of uranium also produces smaller elements, such as krypton and barium.

5 Gamma radiation
Fission reactions release extremely high-frequency ionising gamma radiation, which must be contained.

6 Propagation
A split uranium atom releases two or three neutrons, which go on to collide with more uranium nuclei.

7 Nuclear chain reaction
Uranium atoms' ability to produce neutrons when they split sets up a chain reaction.

8 Energy
Each fission event releases some 200 million electron volts of energy.

Atomic vapour laser isotope separation (AVLIS) in action

Mining uranium

Uranium is a commonly found element in the Earth's crust and ore deposits are distributed across the globe, with the highest concentrations in Australia, Kazakhstan and Canada.

There are three main methods for retrieving uranium. It can be extracted using open-pit mining or underground mine shafts. In pit mines, the ore is exposed by drilling and blasting before being cut directly from the ground, while underground shafts are cut for ore buried deeper down.

The third alternative is in-situ leaching, a form of chemical mining. It uses an acid or alkali to dissolve uranium in the ore into a solution, before pumping the fluid to the surface for processing.

Due to its radioactive properties, mining uranium is a dangerous pursuit. As uranium decays it produces radium, which in turn decays to form carcinogenic radon gas. Inhaled radon gas increases the risk of lung cancer. As a result, advanced ventilation systems are installed in uranium mines and workers wear protective gear.

Uranium has been at the heart of nuclear power for decades, but new materials and techniques are now emerging

What is plutonium?

The 94th element in the periodic table is one of the most dangerous on the planet

Plutonium is radioactive, but it doesn't seem that harmful at first glance. It looks like any other metal, with a silvery sheen that turns dull in contact with the air. Queen Elizabeth II held a piece during a visit to Britain's Atomic Energy Research Establishment at Harwell in 1957. It was warm to the touch, but it didn't hurt. Even eating it doesn't really do any harm, although it's definitely not recommended. Plutonium atoms fall apart by a process called alpha decay. They release particles made from two neutrons and two protons – essentially a helium nucleus. They're so bulky that they can't pass through human skin. Left to its own devices, plutonium decays slowly. The real danger comes when humans interfere.

An isotope of plutonium named plutonium-239 is fissile, which means it can start a nuclear chain reaction. The United Nations lists it as one of only two radioactive isotopes used to make atomic bombs. When you hit a plutonium-239 atom with a neutron, more neutrons will spill out. When those neutrons hit other plutonium-239 atoms, the same thing happens again… and again and again. This releases an enormous amount of energy. Under the right conditions, the energy from nuclear fission can be harnessed for good. The heat released can boil water to make steam, which can spin turbines. A third of the energy made by nuclear power plants comes from plutonium. But it can also be used as a weapon. In 1945, the United States developed a bomb containing a small sphere of plutonium and then used it to devastate the Japanese city of Nagasaki.

The International Panel on Fissile Materials estimates that there are now 140 tonnes of weapons-grade plutonium worldwide. It forms part of the so-called 'nuclear deterrent' in countries like the US, the UK and Russia. You only need a bowling-ball-sized amount of plutonium to make a pit – the core of an atom bomb. But getting your hands on the raw materials isn't easy. Plutonium doesn't really exist in nature; it only comes from nuclear reactors. There's one place in Africa – Oklo in Gabon – where conditions were just right for the element to form naturally, but most plutonium is completely human made.

This is a ring of weapons-grade plutoniur made at Los Alamos National Laborator

Plutonium power

A radioisotope thermoelectric generator (RTG) is a nuclear battery. It captures the heat from decaying plutonium and turns it into electricity using wires called thermocouples. When one end of a thermocouple gets hot, a current starts to flow. RTGs are used to bring power to remote places, from lighthouses on hazardous coastlines to probes in space. But you can't use any old radioactive element to make them. NASA has a strict list of criteria for nuclear batteries. The fuel must be safe in case of a launch accident. It mustn't release too much beta, gamma or neutron radiation because they interfere with equipment. It must be stable, as you don't want it to explode. It must have a long half-life, since you can't change a battery in space. And it must be light, because it needs to get off the launchpad. Of the 2,900 known radioactive isotopes, plutonium-238 is one of just 22 suitable for the job.

The heat from this glowin pellet of plutonium oxic can power a spacecra

he discovery of plutonium

tonium is a relatively new addition to the periodic table. It was first countered during World War II. In 1940, scientists at the Berkeley diation Laboratory were experimenting with uranium, firing heavy drogen atoms at the element to see what would happen. Uranium adioactive, and the bombardment caused it to split apart, easing another radioactive element, neptunium. That element was stable, and it decayed into yet another unknown radioactive ement. Following the planetary theme, the scientists called it itonium. Its potential as a nuclear weapon was immediately ident, and within two years top-secret plutonium production d started at the Metallurgical Laboratory in Chicago.

The X-10 reactor turned uranium into plutonium as part of the Manhattan Project

Oak Ridge National Laboratory in Tennessee makes plutonium for deep-space missions

Did you know?
Plutonium is one of the 15 actinide elements

4 Uses for plutonium

1 Nuclear weapons

Just one kilogram of plutonium can produce a blast equivalent to 10,000 tonnes of chemical explosives. Nuclear treaties aim to prevent the use of these devastating weapons.

2 Space heaters

The electrical components on spacecraft operate best at certain temperatures. Plutonium-238 releases a lot of heat when it decays, helping warm up these delicate instruments in deep space.

3 Thermal energy

The heat generated by decaying plutonium can be captured and turned into electricity. One kilogram of plutonium can produce 8 million kilowatt hours of electricity, making it very efficient, and with no direct carbon dioxide emissions.

4 Nuclear pacemakers

A handful of people in the United States have plutonium-powered pacemakers. These devices aren't dangerous inside the body, but Los Alamos National Laboratory disposes of them carefully when they're removed.

When bismuth metal reacts with oxygen, it forms a colourful, shiny outer covering

Bismuth: the pretty impatient metal

Its low melting point allows easy crystal formation – and can make spoons melt in hot tea

We know about bismuth's unusual beauty because it also melts unusually easily. This metal element turns from solid to liquid at 271 degrees Celsius. That's much lower than other metals, which melt at thousands of degrees Celsius.

As it cools down, highly pure bismuth can form into stepped pyramids and other strange forms. This type of crystal is known as a 'hopper'. The hopper crystal forms because as it cools, bismuth is driven to rapidly form hard edges in preparation for making a cube-shaped crystal. However, the driving force never lets the faces grow. Instead it impatiently forms more edges, leaving metal spirals that merge into unusual structures. The bare metal can then also react with oxygen in the air, forming a shiny, coloured surface.

The low melting point of bismuth can be useful. Mixing bismuth with other metals lowers their melting points, making easy-to-shape alloys. Dentists use such alloys to make moulds of people's mouths. Once people mixed bismuth with antimony (a metal-like element) to make type metal, which was used to print old books.

In the 1800s, Victorian pranksters made spoons from an alloy of eight parts bismuth, five parts lead and three parts tin. The alloy's melting point was low enough for the spoons to vanish into a cup of hot tea. Given that today we know that lead can be poisonous, hopefully the victims didn't finish their drinks!

How to make bismuth crystals

Thanks to its low melting point, you can in theory make pretty bismuth crystals at home. However, the temperatures and open flames involved still mean that great care is needed. Preferably, you should do it with an expert in a lab or a metalsmith's forge.

To form the beautiful crystals, first put high-purity bismuth in an iron or steel container, for example a clean steel can. Then, heat the container to bismuth's melting point, 271°C, using a gas flame, like a stove or a Bunsen burner, or an electrical hotplate. Turn off the heat, and when the bismuth freezes over, poke two holes in the solid surface. Pour the remaining liquid bismuth into a second iron or steel container, and leave to cool.

When it has cooled, break open the metal to find shiny, stepped bismuth crystals. And if you aren't satisfied, you can re-melt the bismuth and try again!

Limescale

Discover what causes limescale and how chemical descalers can remove it

Limescale is a crusty, off-white deposit of calcium carbonate left behind by hard water, and mostly found in your kettle and boiler. Hard water contains an above average concentration of dissolved minerals, like calcium and magnesium ions. Limescale becomes an issue when hard water is heated or left to stand. When the moisture aporates, the dissolved minerals solidify and turn to limescale, which difficult to remove. A build-up of limescale in central heating pipes, ishing machines and dishwashers can restrict water flow, making them s efficient and can even clog them up completely.

Fortunately, there are ways to remove limescale. For example, ion change columns contain ion exchange resin in the form of beads. As iter passes through the column, the calcium and magnesium ions in the iter swap places with the ions in the column, gradually removing the wanted minerals. Citric acid, sesquicarbonate and other compounds 1 also be found in water-softening agents, which prevent the calcium d magnesium ions from forming limescale deposits. Acids are used in ost limescale removers because they react with limescale and produce uble metal salts that can be washed away.

Stainless steel

How does this everyday super-material fight the persistent forces of corrosion?

Stainless steel is an iron alloy with a minimum of 10.5 per cent chromium. Since its discovery in 1913 by metallurgist Harry Brearley, stainless steel has played an essential role across multiple industries. From medical instruments to street furniture, pots and pans and even chemical tankers, stainless steel is an incredibly useful material, saving ictures from rust or corrosion. But since it is comprised mainly of iron, y doesn't this metallic marvel rust?

Rust is the result of iron oxidation, either from the air or from oxygen in ter. The added element of chromium reacts with invading oxygen, acting as at is known as a passive layer, protecting the alloy's surface. This protection urther supported by the inclusion of other elemental sidekicks in the iron y, such as nickel, nitrogen, molybdenum, manganese and silicon. Even so, nless steel does have a weakness; it is susceptible to chlorides found in salt ter, which can break through the protective layer.

Since it in comprised mainly of iron, ıy doesn't this metallic marvel rust?"

Stainless steel is strong but has a high ductility, so it can bend without breaking

Alloys: cutting-edge materials

How mixing metals has given humans tools to help shape the modern world

If you ever pick up an object made of or containing metal, more often than not it's an alloy. This underlines the little-known fact that alloys have been vital in making our lives so comfortable.

Alloys are mixtures of chemical elements where at least one of the elements is a metal. Unlike the atoms in molecules like water, usually they are not chemically bonded together. Instead they just sit side by side. To mix metal atoms together in this way, someone must melt at least one of the metal components, which usually only happens at very high temperatures. That means alloys can be quite difficult to make – but the effort is worth it. Mixing two metals together can improve greatly on what each one can do on its own.

For example, most pure metals are fairly soft, and therefore no good for use in hard-wearing tools like hammers or knives. That's because the layers of atoms that form them can easily slide over each other. Mixing up atoms of different sizes can stop the sliding, as smaller atoms struggle to get around bigger ones. The smaller atoms slot into the gaps between the bigger atoms, making 'interstitial' alloys.

Sometimes different elements being mixed have similar atomic sizes, and one element just swaps in for atoms of the other type. These are called 'substitution' alloys.

Today we have a lot of useful alloys, but there may still be others to find. With more than 80 metals in the periodic table of elements – and scope to mix in non-metal elements – there are a lot of possibilities!

From meteors to spy planes

The first alloys that humans turned into tools and weapons came from space; people used iron and nickel mixtures brought to Earth's surface by meteors. Then, around 2500 BCE, someone mixed copper with tin to create bronze. The resulting alloy was much harder than its ingredients, and gave its name to the Bronze Age.

Since then we've invented even more alloys. In the 19th century removing impurities from steel made it reliably strong, starting the Industrial Revolution. After World War II, Russia and the US wanted faster, lighter spy planes. That led to scientists developing low-weight, high-strength titanium alloys.

The Lockheed Blackbird SR-71 was a stealth aircraft built with lightweight titanium alloys

Pure metal versus metal alloy

Metals mixed together can be more useful than they are on their own

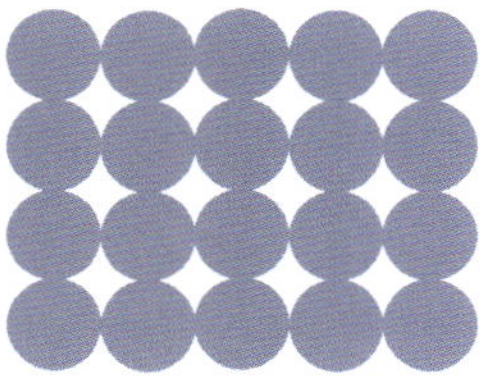

Pure metal

SOFTNESS

Metals like gold and silver are relatively soft, aking them easy to shape into jewellery, but no good for tools.

LOW TENSILE STRENGTH

uminium foil is easy to tear because it has low nsile strength. To improve this, it can be alloyed with magnesium or copper.

MAGNETIC PROPERTIES

Many metals are magnetic, but not always in useful ways. Mixing metals as alloys can improve that.

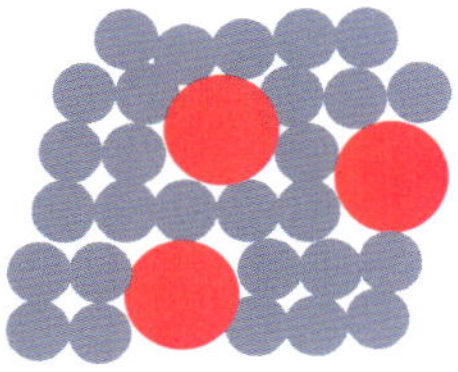

Metal alloy

DUCTILITY

lloys can be more ductile than the metal they're ade from, meaning that it's easier to bend them and draw them out into long wires.

HARDNESS

ecause the atoms in alloys don't slide around as asily, they don't move as much when something hits them.

CORROSION RESISTANCE

n rusts easily, but mixing it with carbon to make eel reduces that problem. Other alloys are even more corrosion resistant.

eteorites provided the first iron alloys ed to make weapons and tools

The world's leading alloys

Alloys are used in many objects in our homes – and sometimes even in our bodies

Amalgam

Dentists mix liquid mercury with a powder containing the other metals to make a soft putty, which quickly hardens to make strong, long-lasting tooth fillings.

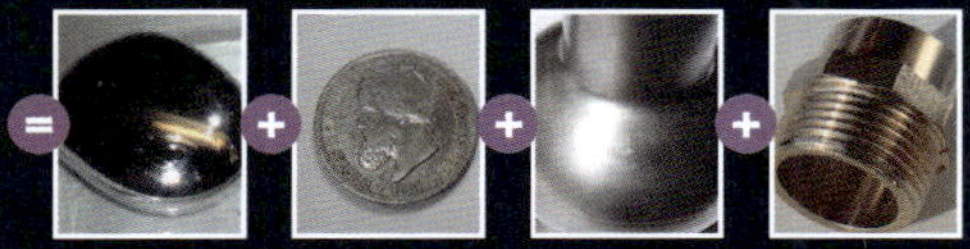

Brass

Brass is a widely used metal alloy because it's easy to form into different shapes. It can be found in products ranging from cars to musical instruments.

Bronze

People first liked bronze because knives made of it stayed sharper than copper ones. It's still used today in heavy-duty machine parts, such as gears.

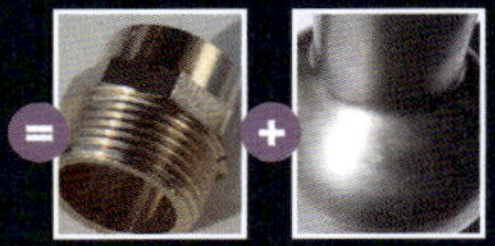

Did you know?
Coins today are generally made from steel – even 'coppers' are steel coated with a tiny amount of copper

Cast iron

With a high carbon content, cast iron melts at a low temperature, and is easy to make into products like kitchen pans and car parts.

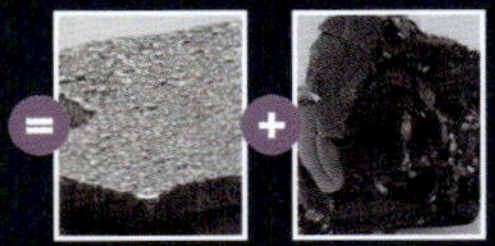

Gunmetal

The name comes from the fact that gunmetal was once used to make cannons. Today it is used for intricate parts often found in plumbing, and also statues.

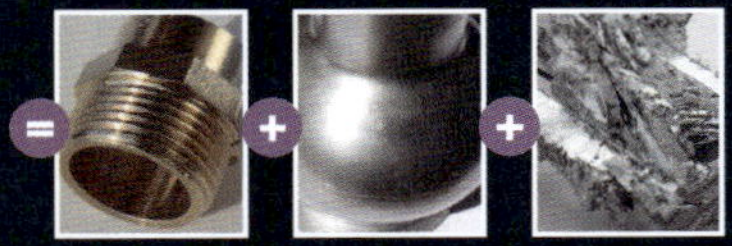

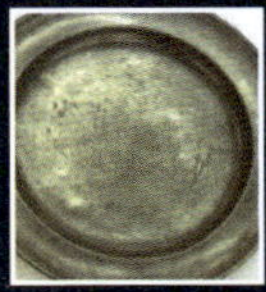

Pewter

Pewter is tin hardened by other elements, but can be shaped at relatively low temperatures. It was used for kitchenware, but today is mostly decorative.

Solder

Solder is mixed to make metals easy to melt. It then hardens as it cools. It joins electrical connections together, and so is very important.

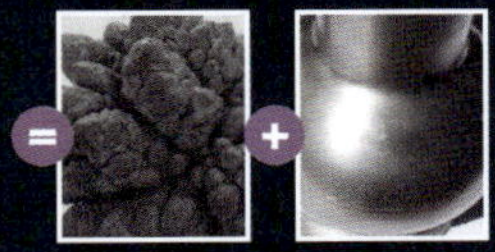

Stainless steel

The chromium in stainless steel forms an outer layer that helps stop the steel rusting. It's common in cutlery and has thousands of other uses.

Steel

Mixing iron with carbon makes it harder, and also reduces the chances of it rusting. Steel is vital in many industries, including building and car manufacturing.

Sterling silver

Pure silver is a very soft metal, so adding 7.5 per cent copper makes it stronger and therefore more suitable for producing practical objects like cutlery.

The power of thermite

When iron and aluminium mix, the consequences can be explosive

Thermite is one of the most dangerous chemical reactions humans have conceived of. Created when aluminium powder is mixed with a metal oxide, typically iron oxide, and ignited, the two metals erupt in a violent burst of light and intense heat. This volatile reaction is the result of oxygen atoms being exchanged between a molecule of iron oxide and aluminium atoms, which means the molecular bonds that grip oxygen to its metal companion are broken and reformed. In doing so, energy is released in the form of heat, in what's known as an exothermic reaction. When reactants such as iron oxide and aluminium have a higher internal energy, known as enthalpy, than the products made by the reaction, the excess energy is released as heat or light.

In the case of a thermite reaction, this excess heat can reach up to 2,500 degrees Celsius, turning the newly separated iron atoms into molten metal. Once the volatile reaction has come to its dramatic conclusion, what remains is a lump of iron, along with some chips of an aluminium-oxide mineral called corundum.

To reap the thermal rewards of thermite, engineers have harnessed its power to weld pieces of metal together. Thermite welding is widely used in metal joining, especially in the railway industry, where thermite is poured over the two ends of rail pieces to fuse them together. Typically, the ingredients for this are bound together with a binding agent – often a carbohydrate called dextrin – to prevent them from separating during the reaction an[d] prolonging the exothermic release.

Did you know?
Thermite typically releases four megajoules of energy per kilogram; gasoline releases 49 megajoules per kilogram

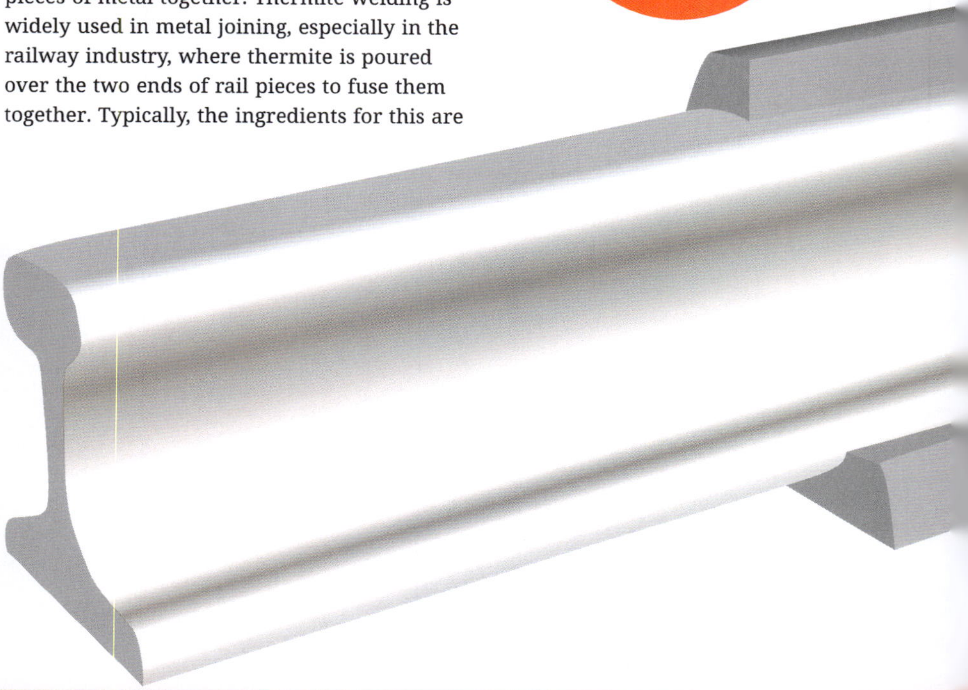

Goldschmidt's discovery

In 1893, German chemist Hans Goldschmidt came across thermite through what later became known as the Goldschmidt process, while investigating ways to extract pure metal from natural ores. At the time, Goldschmidt relied on the traditional method of obtaining iron from its ore by heating it with carbon. During this process, the carbon combines with oxygen in the iron oxide ore, releasing carbon dioxide as a byproduct and leaving the pure iron behind. However, this method of oxygen stripping can leave carbon contaminants within the iron. It was while Goldschmidt was experimenting with aluminium as a possible alternative to carbon that he uncovered thermite's impressive exothermic reaction. Seeing its potential to be used in metal joining, Goldschmidt patented the welding compound thermite.

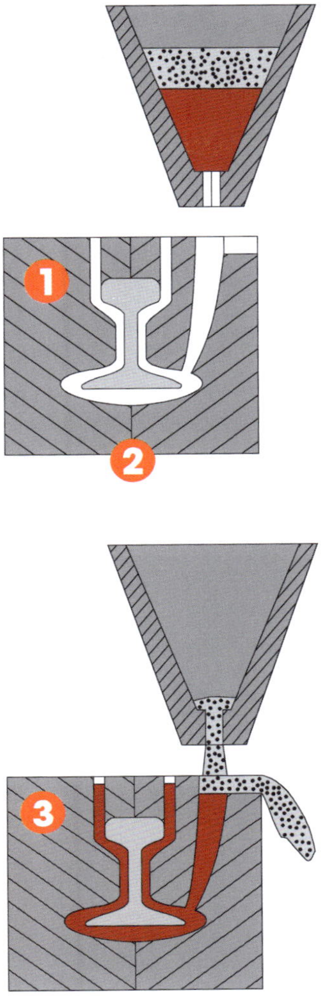

Metal joining

How thermite is used to repair train tracks and steel pipes

1 Mould
A mould is used to snugly fit over the shape and size of the tracks that will be joined together.

2 Clamped together
The mould clamps the two tracks together, leaving a slight gap for the thermite to be added.

3 Preheating
Before the thermite is used, the gap between the tracks is preheated to reduce the amount of shrinkage on the metal when it cools down.

4 Melting pot
The thermite crucible is made of hardy materials such as graphite, which can handle reaction temperatures of around 3,000 degrees Celsius.

5 Thermite mixture
The ratio of the thermite powder is typically one part aluminium and three parts iron oxide.

6 Igniting powder
Following ignition, the thermite reaction begins producing molten iron to weld the tracks together.

7 Aluminium slag
As a byproduct of the reaction, the less dense aluminium oxide floats to the top of the crucible.

Did you know?
The thermite welding process was patented in 1899

Lumps of iron are created after a thermite reaction

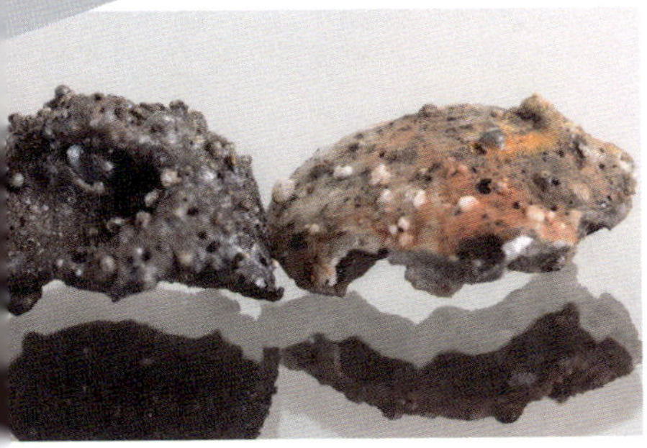

Thermite welding in action on a railway track in Dublin

Bombs away

The explosive nature of thermite makes it a suitable ingredient for making incendiary bombs. What's particularly devastating about thermite bombs is their ability to rain down molten metal. Following the ignition of the thermite reaction by a fuse, the burning metals erupt in a firework-like explosion, maintaining thousand-degree temperatures as they fall to the ground. Not long after the discovery of thermite, the German military enlisted it as a weapon. In 1915, during World War I, thermite bombs were deployed by German forces in Zeppelin raids on England. During World War II, thermite bombs got an upgrade with more energetic elements, such as copper and magnesium, to maximise the damage they could cause. The use of thermite bombs is now regulated under Protocol III of the UN Convention on Certain Conventional Weapons.

A thermite explosion at a Russian arms exhibition in 2013

Killer chemicals

Nature has many gruesomely deadly substances, but understanding them keeps us safer

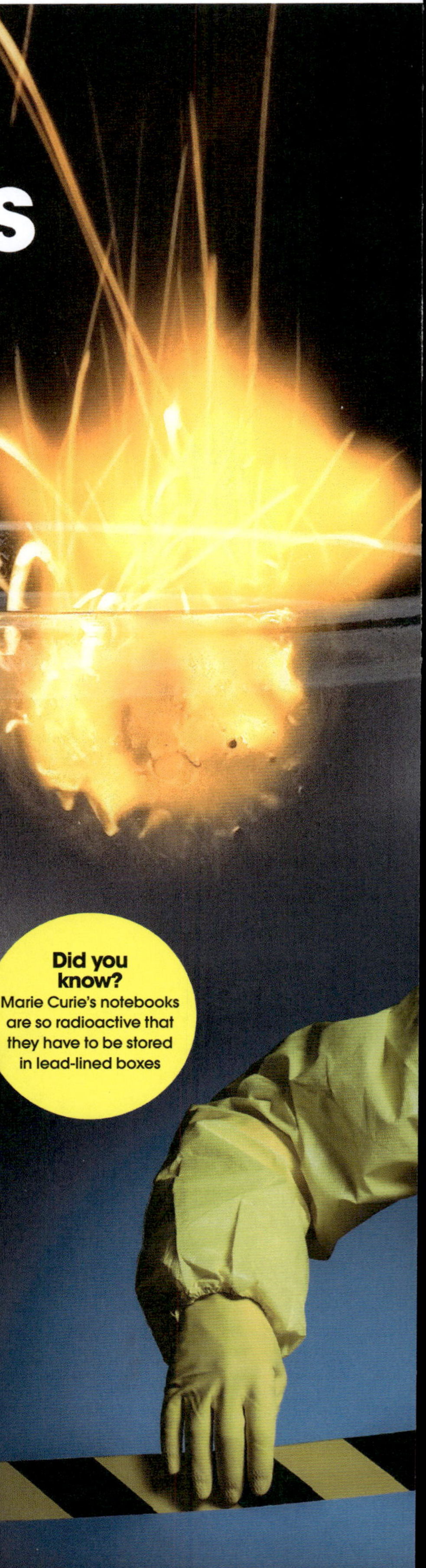

Marie Curie is one of few women to win science's highest honour, a Nobel Prize, but her work also brought her an early death. She and her husband Pierre spent years carefully grinding radioactive rocks and soaking the powder in acid. In 1898 they discovered the new element polonium, and a few months later, Marie would go on to discover radium too. But the Curies didn't know how risky their work was. Scientists had discovered radioactivity just a few years earlier. We now know that radioactivity is very dangerous. Worse still, polonium may be the most radioactive element.

The danger comes from the building blocks forming elements' atomic structures. At each atom's centre is its nucleus. In radioactive elements, the nucleus breaks apart, firing out smaller chunks that can harm people. Also, a cloud of electrons surrounds each atom's nucleus. The number of electrons an atom has and how the nucleus is built make all elements behave differently. From this chemistry emerge many different hazards.

Some hazards arise because elements are toxic or poisonous. Most often, such poisons work because two different elements are similar. We need many essential elements to live, powering the biological machines in our cells. Therefore we must take them in when we eat and drink. But consuming elements that mimic the essential ones stops the biological machines working.

Other hazards come from elements' reactivity. Usually this happens when elements strongly give electrons to other atoms, or take them away. If the chemical reactions this causes are fierce enough, they become fires or explosions. Yet these same properties mean that the substances produced are then very safe and stable. Chlorine's reactivity makes it poisonous, and sodium's makes it explosive in water. But together they make sodium chloride, which is so safe we eat it as table salt. And when atoms assemble into molecules like nitroglycerin, these properties can be even more complex.

But why would anyone want to understand these dangerous chemicals? It's not because scientists want to poison or kill people. Instead, knowing about these substances' bad effects helps everyone avoid being hurt.

Did you know?

Marie Curie's notebooks are so radioactive that they have to be stored in lead-lined boxes

The explosive that heals hearts

Nitroglycerine was the first explosive to be more powerful than gunpowder. Like many other explosives, such as trinitrotoluene (TNT), its power comes partly from the atoms responsible for the 'nitro' in its name. Nitro groups link together one nitrogen atom and two oxygen atoms. Nitroglycerine therefore has lots of oxygen atoms, along with carbon, hydrogen and nitrogen atoms. When it burns the high oxygen content means the atoms react explosively, as they seek to produce more stable substances.

Nitro groups also play important roles in the body. They break down to form chemical signals that tell blood vessels to widen. Nitroglycerine is used to help reduce the pain people experience when they have problems with their heart.

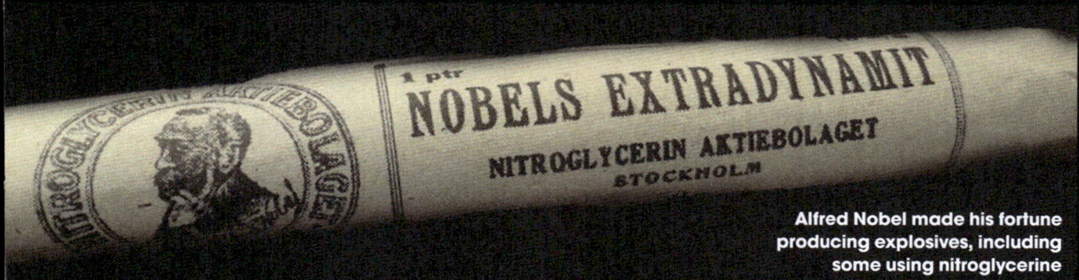

Alfred Nobel made his fortune producing explosives, including some using nitroglycerine

The element that killed a Russian spy

In November 2006, former Russian secret agent Alexander Litvinenko became the first-ever person that we know to have been killed by polonium poisoning. This horrible death shows just how powerful his enemies were, in part down to the extreme rarity of the poison. Polonium itself is a rare element, found in rocks that also contain uranium. A gram of such a rock contains around one ten-billionth of a gram of polonium.

There are also 42 different forms of the element, known as isotopes, all of which give off dangerous radiation. Litvinenko died from the powerful radiation of the polonium-210 isotope disrupting the working of his cells. He probably drank the substance three weeks earlier, in the form of dissolved polonium chloride. The radioactive polonium could then quickly pass from Litvinenko's stomach to his blood, and then to the rest of his body.

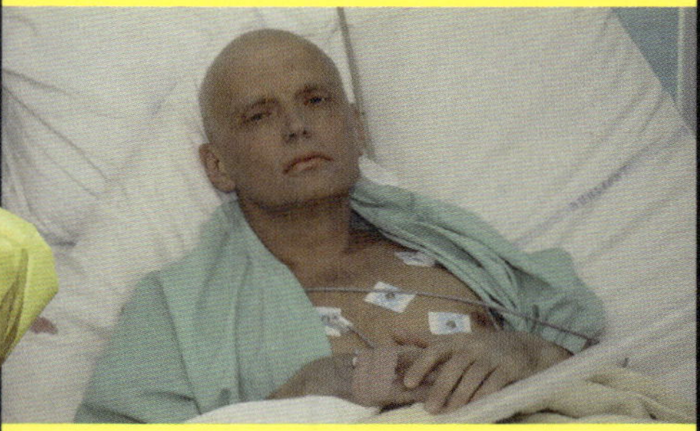

Litvinenko was poisoned with polonium-210, which affects rapidly dividing cells the worst

Explosives on your table?

Sodium chloride is simple table salt, but has two very dangerous ingredients

Sodium (Na)

You rarely see sodium as a metal element because it's so reactive. This silvery metal is soft enough to cut with a sharp knife. But don't be deceived by its softness – drop it in water and it will explode!

Chlorine (Cl)

You really don't want to see the yellow-green gas that is pure chlorine, because breathing in very high levels can kill you. Chlorine's hunger for electrons means that tiny amounts can kill off bacteria.

Flashy but familiar chemistry

One way to make sodium chloride is dropping sodium metal into chlorine gas. Perhaps adding a little water, there would be a bright flash of a violent reaction. The white smoke that remains afterwards is much like table salt.

Table salt

In the flashy reaction, chlorine atoms steal electrons from sodium metal. The resulting sodium and chloride ions nestle together to produce crystals of table salt. These ions are very stable, so don't react when eaten.

10 Deadly elements

Hydrogen Flammable

If one-fifth of the air we breathe wasn't made up of oxygen, hydrogen might be relatively safe. But hydrogen and oxygen react rapidly, burning together to make water.

Hydrogen *** H 1.008 1									
Lithium * Li 6.941 3	Beryllium * Be 9.012 4								
Sodium * Na 22.99 11	Magnesium * Mg 24.31 12								
Potassium * K 39.10 19	Calcium * Ca 40.08 20	Scandium * Sc 44.96 21	Titanium * Ti 47.87 22	Vanadium * V 50.94 23	Chromium * Cr 52.00 24	Manganese * Mn 54.94 25	Iron * Fe 55.84 26	Cobalt * Co 58.93 27	Nic * 58.
Rubidium * Rb 85.47 37	Strontium * Sr 87.62 38	Yttrium * Y 88.91 39	Zirconium * Zr 91.22 40	Niobium * Nb 92.91 41	Molybdenum * Mo 95.94 42	Technetium * Tc [98] 43	Ruthenium * Ru 101.07 44	Rhodium * Rh 102.91 45	Pa * 10
Caesium * Cs 132.91 55	Barium * Ba 137.33 56	LANTHANIDES	Hafnium * Hf 178.49 72	Tantalum * Ta 180.95 73	Tungsten * W 183.84 74	Rhenium * Re 186.21 75	Osmium * Os 190.23 76	Iridium * Ir 192.22 77	Pla * 19
Francium * Fr [223] 87	Radium * Ra [226] 88	ACTINIDES	Rutherfordium **** Rf [267] 104	Dubnium **** Db [268] 105	Seaborgium **** Sg [269] 106	Bohrium **** Bh [270] 107	Hassium **** Hs [269] 108	Meitnerium **** Mt [278] 109	Da ** [2

Lanthanum * La 138.91 57	Cerium * Ce 140.12 58	Praseodymium * Pr 140.91 59	Neodymium * Nd 144.24 60	Promethium * Pm [145] 61	Samarium * Sm 150.36 62	Europium * Eu 151.96 63	Ga * 15
Actinium * Ac [227] 89	Thorium * Th 232.04 90	Protactinium * Pa 231.04 91	Uranium * U 238.03 92	Neptunium * Np [237] 93	Plutonium * Pu [244] 94	Americium * Am [243] 95	Cu * [2

Chromium Carcinogen

Chromium metal that has reacted wi air is stable and shiny, so is used in c parts. But removing six electrons fron metal chromium atom gives cancer-causing substances.

Caesium Explosive/flammable

If you ever come across caesium, handle it with care! Left alone in air, caesium metal bursts into flames. Put in water, it explodes violently.

Plutonium Radioactive

Plutonium's atomic structure enables nuclear weapons, and makes it highly radioactive. Being radioactive means that its atoms break down, spitting out chunks that can badly damage human cells.

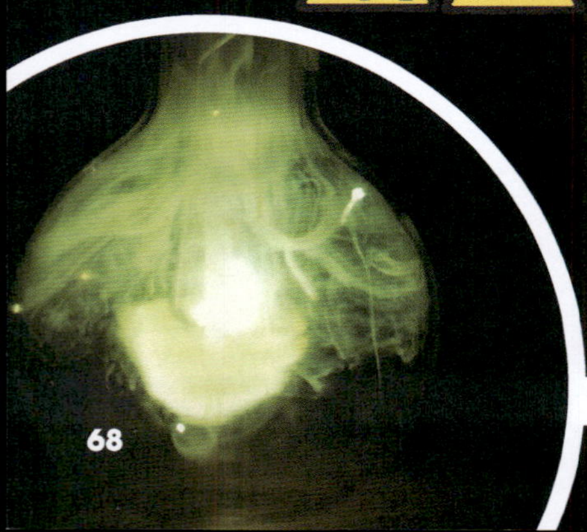

Sodium metal in chlorine gas reacts with a bright flash to make salt

Element state

* solid ** liquid *** gas **** unknown

Element group

- Alkali metal
- Alkaline earth metal
- Lanthanide
- Actinide
- Transition metal
- Post-transition metal
- Metalloid
- Other non-metal
- Halogen
- Noble gas
- Unknown chemical properties

Arsenic Poison
The well-known poison arsenic also confuses our cells. That's because its properties are similar to the essential element phosphorus, which is just above it in the periodic table.

Fluorine Corrosive
Fluorine's atomic structure makes it greedy for electrons from other atoms and therefore extremely reactive. It can corrode, degrade or modify nearly everything, making it very toxic and corrosive.

Antimony Poison
Antimony is below arsenic in the periodic table, and the two elements have similar poisoning effects. This is because both behave like phosphorus. Antimony damages people's livers and makes them vomit.

admium Poison
true heavy metal, cadmium can terfere with how calcium builds up ones in our bodies. Cadmium can so mimic zinc, which we need to e, but without its benefits.

Thallium Poison
Thallium is physically similar to another element, potassium, which keeps us alive. But thallium doesn't behave exactly the same way, so it confuses our cells and stops them working.

Lead Poison
Lead used to be very common in paint and petrol. However, we now know it damages the brain and kidneys. Today it is used much more carefully.

5 Facts
Sometimes dangerous, sometimes useful

1 Painting a gloomy picture
Dutch artist Vincent van Gogh wanted to eat paint to try to kill himself. The yellow paint he used for his famous *Sunflower* pictures contained chromium, making it poisonous.

2 Hydrogen fuels cleaner heating
The natural gas many people use for heating their homes produces greenhouse gases that cause climate change when it burns. Because hydrogen burns easily, it might replace natural gas.

3 Friendly in low doses
To keep swimming pools clean without harming ourselves, we use very small amounts of chlorine. Usually swimming pools contain just one molecule of chlorine for every million water molecules.

4 Exceptionally bright hopes
Cadmium's poisonous nature means companies are removing it from most products, but there is one exception: some solar panels are made from a cadmium compound whose benefits outweigh their risks.

5 Drop-dead gorgeous
Poisonous antimony compounds were used in ancient Egypt as eye make-up in around 3100 BCE. The wearers had no idea just how unhealthy their vanity was!

Toxic science

Discover some of the deadliest substances known to humankind

Toxic substances include anything that can physically harm us after we inhale, swallow or touch it, from an innocent bee sting to full-blown cyanide poisoning. Defining toxicity is tricky since almost anything is poisonous at high enough doses – even water. Acute poisoning follows just one exposure, for example, nibbling a death cap mushroom, but chronic exposure – like inhaling cigarette smoke over decades – can be equally, if not more, damaging.

Toxins are toxic substances produced by living organisms. These organisms use toxins mainly to ward off predators or paralyse prey. Small but deadly bacteria produce some of the most potent toxins known, including botulinum toxin A (Botox). Other toxic substances occur naturally on Earth, such as the hydrogen sulphide produced by volcanic eruptions. We humans have even invented toxins for use as pesticides, insecticides (eg DDT) or chemical weapons (eg sarin, VX).

Targeting different parts of the body, toxic substances can damage us in an alarming number of ways. Neurotoxins are some of the most effective, affecting the brain and nervous system and causing muscles to freeze or twitch uncontrollably. Other substances can burst our red blood cells or cause allergic reactions.

But not everyone is affected by toxic substances in the same way. How toxic a chemical is depends on how easily it is absorbed, metabolised and eventually expelled by the body. Children are generally more vulnerable as their bodies are not able to get rid of toxic substances as effectively. Different species are also more or less susceptible to various poisons – for example, it takes 1,000 times more dioxin to kill a hamster than it does to kill a guinea pig.

KEY

Toxicity: 1 = Unlikely to kill **5** = Super-deadly
Rarity: 1 = Very common **5** = Very rare

Botulinum toxin A (Botox)

This is the most toxic substance in nature: just one gram (0.04 ounces) could kill 14,000 people if swallowed – or 8.3 million if injected! Produced by Clostridium botulinum bacteria, this neurotoxin is responsible for botulism, a rare but life-threatening illness transmitted principally through contaminated canned food. Botulinum disrupts communication between nerves and muscle cells, gradually paralysing its victims and finally leading to respiratory failure. Extremely small doses of botulinum toxin can, however, be used to treat muscle spasms and excessive sweating and to paralyse the muscles that cause wrinkles (sold commercially as Botox).

THE STATISTICS...

Main symptoms: Double vision, droopy eyelids, difficulty swallowing, slurred speech, muscle weakness, paralysis

Antidote: Horse-derived antitoxin

Time to death: Rarely fatal when treated

Toxicity rating: 5

Rarity rating: 4

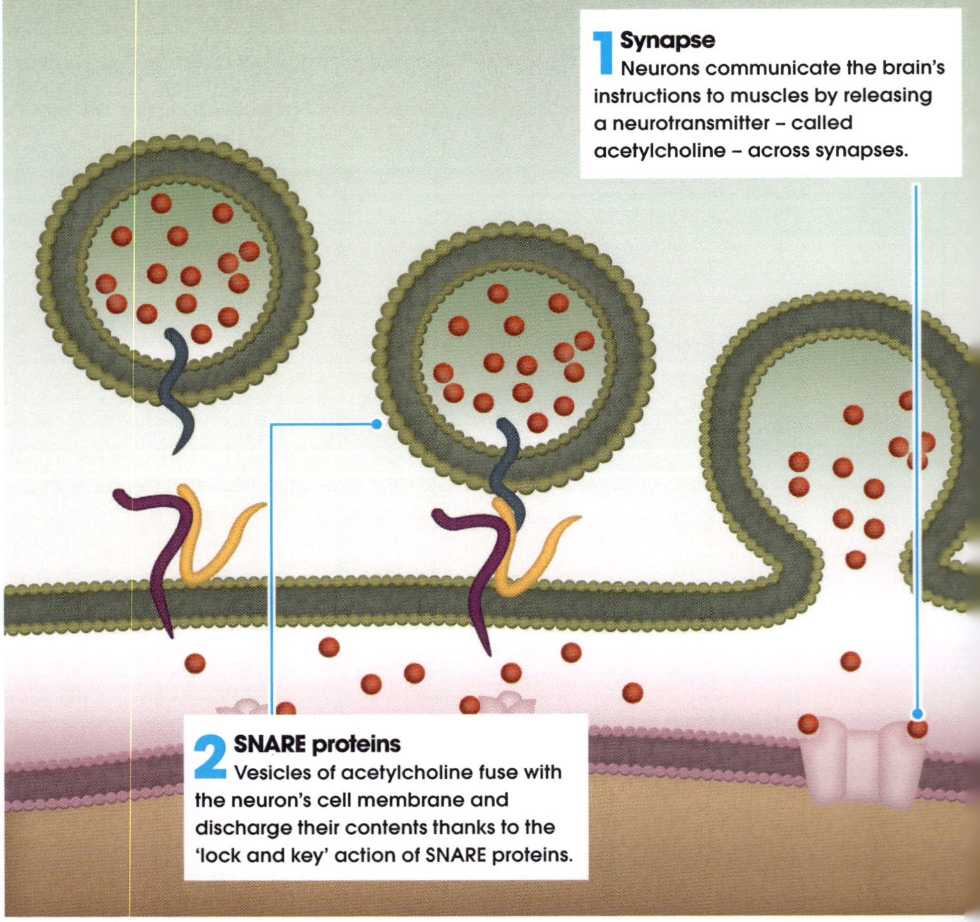

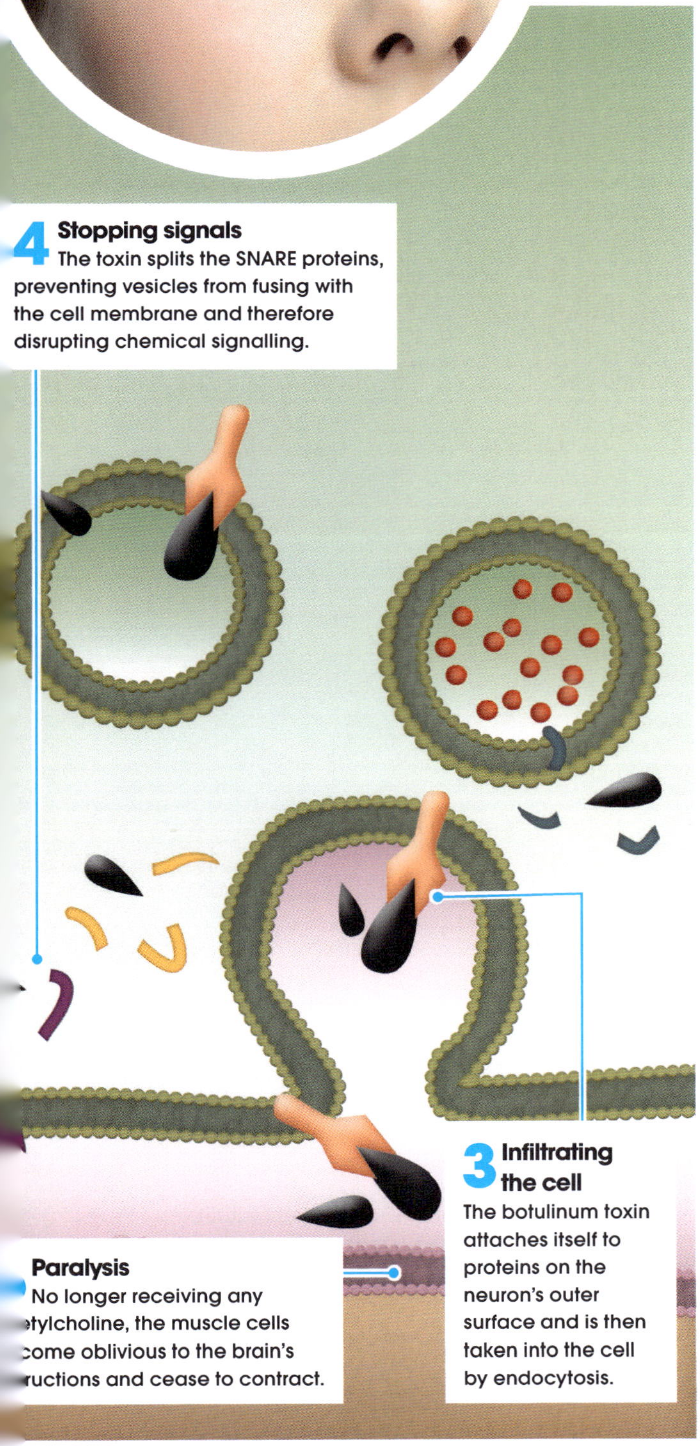

Asbestos

Asbestos is the name given to a handful of different minerals that share one common feature: bunches of fibrous crystals. Boasting an array of insulating properties topped off with a low price tag, asbestos was a popular building material until its toxic effects came to light. With repeated inhalation, asbestos fibres accumulate in the lungs, causing deadly diseases like asbestosis, an inflammatory lung condition, and cancer. These diseases typically don't develop until 15-30 years after exposure. Although now banned in most countries, older buildings can still release the harmful crystals when demolished.

THE STATISTICS...

Main symptoms: Shortness of breath, coughing, chest pain

Antidote: No current cure for asbestosis, but relief treatment

Time to death: Various

Toxicity rating: 4

Rarity rating: 3

Ricin

Found in the castor oil plant, ricin is a toxic protein that wreaks havoc on ribosomes, the cell's protein builders. The result is severe damage to major organs. Just one milligram of ricin is enough to kill an adult if inhaled or ingested, leading many countries to investigate its use as a biological weapon. The castor oil plant's popularity as an ornamental shrub and the relative ease of extracting the toxin from castor beans have also made ricin the poison of choice for many assassins.

THE STATISTICS...

Main symptoms: Diarrhoea, nausea, accelerated heart beat, hypotension, seizures

Antidote: The UK military has developed an antidote, but it remains to be tested on humans

Time to death: 2–5 days

Toxicity rating: 5

Rarity rating: 2

Carbon monoxide

Colourless and odourless, carbon monoxide gas has a knack for going unnoticed. It is produced by the incomplete combustion of organic fuels including gas, coal and wood – occurring, for example, when inadequate ventilation deprives a gas-burning stove of oxygen. As a result, carbon monoxide poisoning is the most common type of air poisoning around the home. Carbon monoxide molecules bind tightly to haemoglobin, the oxygen-carrying protein in blood. Taking oxygen's place, they prevent blood from delivering oxygen to cells. You can reduce the risk by keeping your home well ventilated and servicing appliances such as boilers every year.

THE STATISTICS...

Main symptoms: Headache, nausea, vomiting, dizziness, fatigue, weakness, loss of consciousness

Antidote: Oxygen

Time to death: 2-3 minutes in acute cases

Toxicity rating: 4

Rarity rating: 1

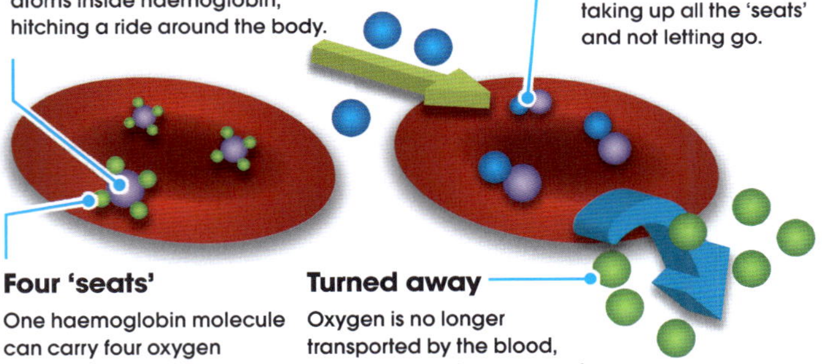

Although toxicity varies across species, there can be enough tetrodotoxin in a single pufferfish to kill 30 people

Toxic household

Keep an eye on the toxic substances lurking in your home...

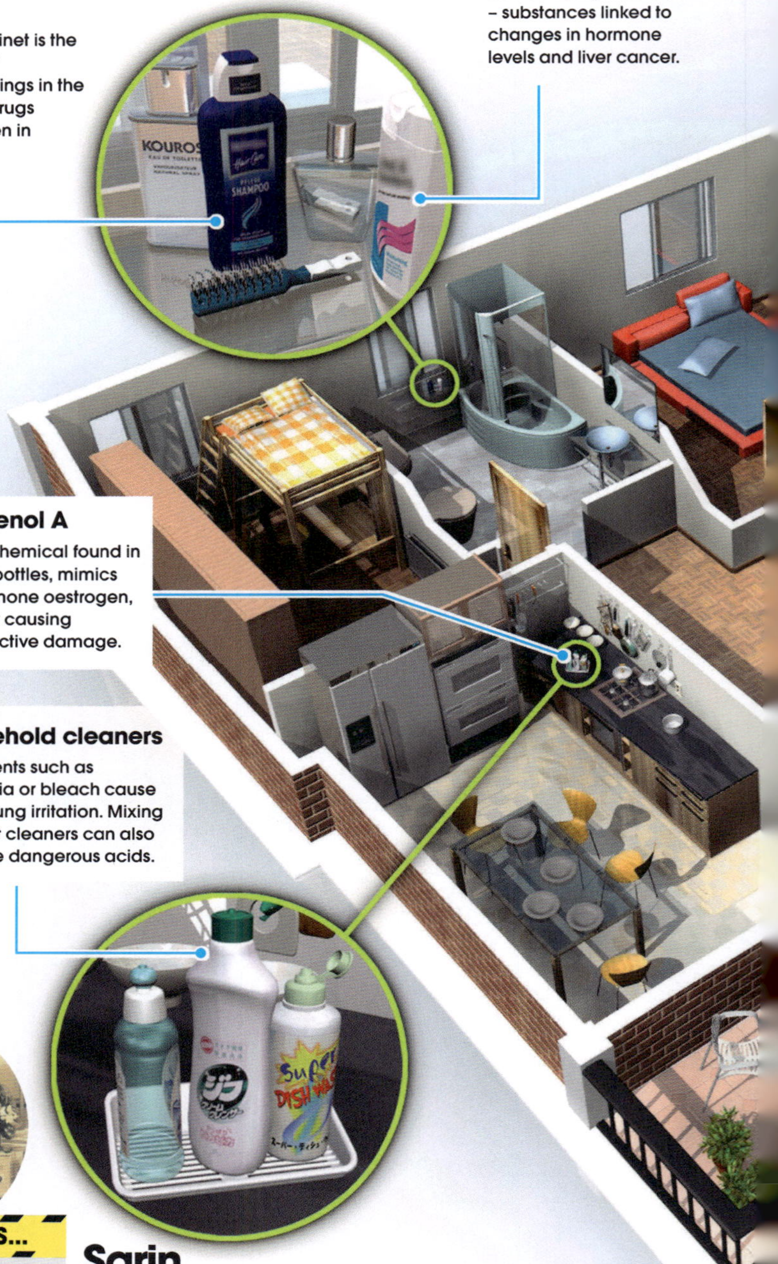

Phthalates

Personal care products, and also vinyl flooring, can contain phthalates – substances linked to changes in hormone levels and liver cancer.

Medicines

The medicine cabinet is the greatest source of accidental poisonings in the home, with most drugs harmful when taken in excessive doses.

Bisphenol A

BPA, a chemical found in plastic bottles, mimics the hormone oestrogen, possibly causing reproductive damage.

Household cleaners

Ingredients such as ammonia or bleach cause skin or lung irritation. Mixing different cleaners can also produce dangerous acids.

Tetrodotoxin

Thrill-seeking Japanese diners are sometimes tempted to try fugu, a variety of pufferfish. The catch? If the chef slips up, they risk being poisoned with tetrodotoxin, a potent neurotoxin contained in the fish's gonads, liver, intestines and skin. Opening nerves' ion channels, tetrodotoxin acts similarly to batrachotoxin to block nerve impulses, causing paralysis and death by respiratory failure. Although chefs need a licence to serve fugu, mishaps still poison an estimated 200 people each year, with half of them dying. Pufferfish are not the only ones to use tetrodotoxin; it is one of the most common toxins in the marine world, employed by scores of fish, crabs and molluscs, including the blue-ringed octopus.

THE STATISTICS...

Main symptoms: Numbness of the lips and tongue, followed by paralysis that spreads to the entire body, heart failure

Antidote: None known

Time to death: 4-6 hours

Toxicity rating: 4

Rarity rating: 4

Cyanide

THE STATISTICS...

Main symptoms: Nausea, rapid breathing, dizziness, headache, convulsions – leading to death

Antidote: In smaller doses hydroxocobalamin is one known antidote, but generally fatal

Time to death: As little as a minute

Toxicity rating: 5

Rarity rating: 2

Whether inhaled or ingested, cyanide is one of the fastest-acting poisons known, sealing death sentences in minutes. Chemically speaking, a cyanide is a compound with a triple bond between a carbon and a nitrogen atom. Hydrogen cyanide gas and solid sodium or potassium cyanide are highly toxic, preventing the body's cells from using oxygen and starving the heart and the brain. Certain fruit pits contain cyanide and small quantities of hydrogen cyanide are present in engine exhaust fumes. Industrial uses include gold mining and pesticides – one of which was used by the Nazis in gas chambers.

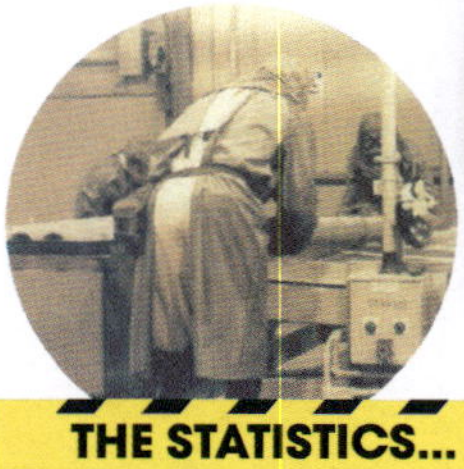

Sarin

THE STATISTICS...

Main symptoms: Constriction of pupils, drooling, difficulty breathing, loss of control over bodily functions, convulsions

Antidote: Atropine

Time to death: 15 minutes to a few hours

Toxicity rating: 5

Rarity rating: 5

Sarin is a human-made nerve agent, first developed as a pesticide by German scientists in 1938. A colourless, tasteless but extremely volatile gas, it works by inhibiting the body's enzyme which breaks down the neurotransmitter acetylcholine, causing it to accumulate at nerve endings. This signals to muscles to contract uncontrollably, triggering a range of unpleasant effects which culminate in death by asphyxiation. Like all chemical weapons, sarin is outlawed and has been used only a handful of times: like during the Iran-Iraq War in the Eighties, and in terrorist attacks on the Tokyo subway in 1995.

Flame retardants

PBDEs (polybrominated diphenyl ethers) found in mattresses and furniture to make them fireproof may cause learning and memory deficits.

VOCs

Just after fitting, the glues and dyes used in new flooring can emit harmful volatile organic compounds (VOCs).

Carbon monoxide

Gas-burning fires can produce potentially deadly carbon monoxide gas if they don't receive enough ventilation.

Lead paint

Houses built before 1978 may contain neurotoxic lead-based paint which can be exposed if it peels.

Cigarettes

Smokers inhale over 700 poisons with each drag, including arsenic, benzene, cadmium, hydrogen cyanide, carbon monoxide and even radioactive polonium-210.

Garden chemicals

xposure to pesticides, herbicides nd fertilisers has been linked to sthma as well as various eurological, developmental and nmunological disorders.

TCDD

TCDD is the deadliest of the dioxins. These chemicals occur in the natural world but are produced in much larger quantities by industry. Dioxins persist for a long time, accumulating in the fat cells of living organisms. As a result, small quantities of dioxins may go unnoticed, but over time they can damage the immune and reproductive systems and increase the likelihood of diabetes and cancer. High doses such as those experienced during the Vietnam War with the USA's use of Agent Orange – a herbicide contaminated with TCDD – spark an immediate reaction. They are also thought to cause cancer and birth defects years later, although TCDD's effect on the body is not yet fully understood.

THE STATISTICS...

Main symptoms: Skin disease (chloracne) and discolouration, lung infection; in the longer term: cancer, birth defects

Antidote: None

Time to death: Unconfirmed

Toxicity rating: 4

Rarity rating: 2

Batrachotoxin

Batrachotoxin is the deadliest ingredient in a lethal cocktail of toxins secreted by certain poison-dart frogs. Native tribes use it as a weapon, dipping their blowgun dart tips in the frogs' toxins – these darts kill prey almost instantaneously. The frogs don't actually produce batrachotoxin themselves but obtain it by eating poisonous beetles. Batrachotoxin opens nerve cells' ion channels permanently, preventing them from creating an electric potential. This blocks cell signalling, paralysing muscles. Heart muscles are particularly sensitive to the toxin, leading to an irregular pulse and, soon after, a heart attack.

THE STATISTICS...

Main symptoms: Convulsions, salivation, muscle contractions

Antidote: None

Time to death: Under 10 minutes

Toxicity rating: 5

Rarity rating: 4

Worst of the rest

1 Alpha-amanitin

This deadly toxin is taken up by the liver, where it inhibits an enzyme needed for cell division, causing liver failure.

Found in: Death cap and destroying angel mushrooms

2 Arsenic

Once believed to have killed Napoleon (which has now been disproved), arsenic disrupts cells' energy transport, leading to organ failure.

Found in: Wood-preserving chemicals, insecticides

3 VX

VX is the most toxic nerve agent ever synthesised – ten times more toxic than sarin.

Found in: Russia and the USA – but now being destroyed

4 Strychnine

This poison causes some gruesome symptoms such as muscle convulsions, arching of the body and facial spasms.

Found in: Strychnos trees, rodent pesticides

5 Polonium-210

If ingested, this radioactive material bombards the body with deadly alpha particles.

Found in: Certain rocks, can settle on broad-leaf plants

Digitalis

Digitalis, or foxglove, owes its toxicity to cardiac to the glycosides digitoxin and digoxin – compounds with the capacity to help and harm. When ingested, glycosides affect the behaviour of heart muscles. In controlled doses, they can regulate the heart beat and treat congestive heart failure. But taking too much digitalis medication, or eating those parts of the plant, can trigger a fatal heart attack; that said, eating foxgloves usually induces vomiting which can prevent an overdose. US serial killer Charles Cullen poisoned at least 29 elderly patients in nursing homes by administering overdoses of insulin and digoxin.

THE STATISTICS...

Main symptoms: Nausea, vomiting, diarrhoea, visual disturbances, hallucinations, reduced heart rate

Antidote: Digoxin Immune Fab: this binds to the toxin, preventing it from acting on the body

Time to death: Rarely fatal but can kill in 24 hours

Toxicity rating: 3

Rarity rating: 2

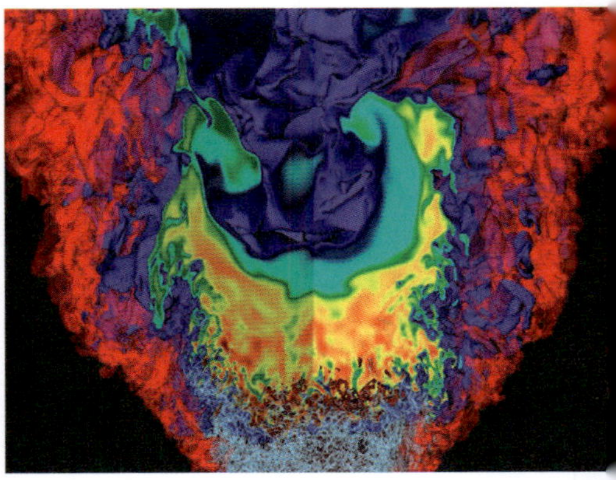

Amazing chemical reactions

Ten of Earth's most awe-inspiring chemical and physical reactions at an atomic level

Chemistry is a fascinating subject. The way atoms behave determines everything from the food we eat and the clothes we wear, to our genetic makeup and how we feel. The beautiful simplicity of the periodic table describes just 118 elements that form every known material – 98 if you discount those that can only exist in a laboratory. These elements react in a variety of ways, forming hundreds of millions of compounds – two or more elements bonded together – giving us the diversity of materials that make up our world.

Some reactions are physical, not chemical, and the way to tell is whether or not there has been a change in the chemical formula. Melting ice is a physical change because water is the same substance as ice, just in a different state of matter. Burning coal, on the other hand, is a chemical change, as coal and oxygen combine to make carbon monoxide – a chemically unique material. A chemical reaction drives the change of one substance into another and the reactions are generally identified by colour changes or a release of energy – often in the form of heat, light and sound: the ingredients of an explosion.

It is impossible to cover the astonishing range of chemical reactions that happen in our universe in one article, so we've picked ten standout ones which have dramatic results.

₹eaction types

›istinguishing between physical and chemical reactions is one thing, yet hemists have identified five common ways that chemical changes can ‑ broken down further. These are: synthesis, decomposition, single ›placement, double replacement and oxidation/reduction (redox).

Some reactions can exhibit characteristics of more than one of these labels – as all chemical reactions are caused by the sharing, gaining or losing of electrons. However it's helpful to categorise reactions by the distinct ways in which they behave.

›ynthesis

synthesis reaction occurs hen two or more chemical ›ements or compounds ›act together to make a ›ore complicated ›mpound. For example burning hydrogen (H_2) and ›xygen (O_2) gas forms the ›ore complex water ›olecule (H_2O).

A + B → A-B

Decomposition

Decomposition is the breaking down of two or more complex molecules into simpler ones. Passing an electric current through water (H_2O), results in the 'decomposition' of the water molecule into its basic elements: hydrogen (H_2) gas and oxygen (O_2) gas.

A-B → A + B

Single replacement

When one element is bumped by another in a compound, it's a single replacement reaction. Reactions with metals and acids often fall into this group. Magnesium (Mg) and hydrochloric acid (HCl) react to form magnesium chloride ($MgCl_2$) and hydrogen (H_2), where Mg replaces H_2.

A-B + C → A-C + B

Double replacement

Sometimes, compounds will 'swap' components – this is called a double replacement reaction. For example, hydrochloric acid (HCl) and sodium hydroxide (NaOH) react together, producing sodium chloride (NaCl) and water (H_2O). In this reaction, the hydrogen and sodium atoms have switched places.

A-B + C-D → A-C + B-D

Redox

Oxidation and reduction (ie redox reactions) describe a chemical change where electrons are transferred. You can't have oxidation (loss of electrons) without reduction (gain of electrons). When H_2 burns with O_2, the H_2 becomes oxidised and the O_2 is reduced.

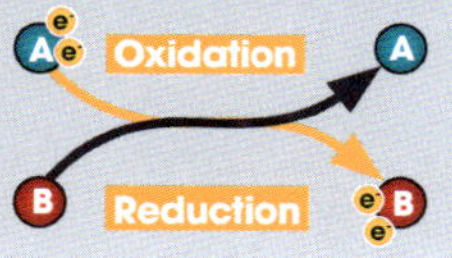

⊤hermite reaction

›eadliness: ☠☠☠☠
›gredients: Aluminium (Al); iron oxide (Fe_2O_3)
›ore process: Single replacement

: In nature ✓ In lab ✖ At home ✖ Toxic

›ermite is a hot reaction that ›nsists of metal powder and a ›etal oxide (usually aluminium ›nd iron oxide); the latter more ›ommonly known as rust. The ›aracteristics of thermite reactions ›entre on their ability to heat very ›nall areas to incredibly high ›mperatures. You don't think of ›etals as burning very easily, but in ›e right conditions – and very high ›nition temperatures – they can.

Thermite reactions are used for ›elding train tracks together and ›mperatures as high as 2,500 ›egrees Celsius can be reached. ›ue to the blazing heat, products of thermite reactions are liquid, making them perfect for welding. As thermite reactions have their own supply of oxygen from the metal oxide they can work even in the absence of air, such as underwater and in space.

Aluminium and iron oxide are heated, often with magnesium ribbon as a fuse, and oxygen from the iron oxide breaks its bond to combine with the aluminium in order to form aluminium oxide and iron. Special face masks that have been fortified with UV protection must be worn when welding due to the intense radiation.

The metal melter

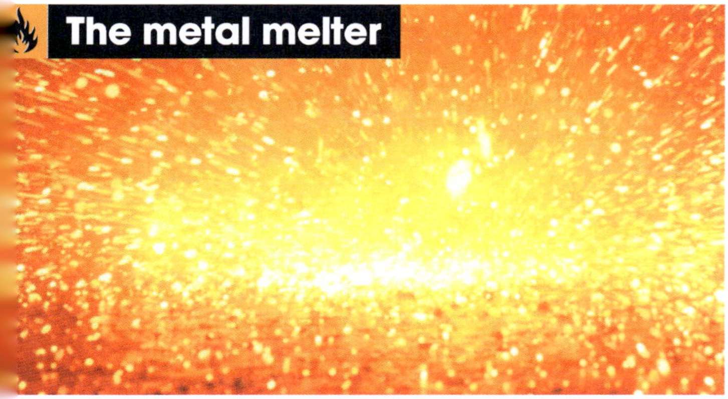

The crystalliser

Copper sulphate crystals

Deadliness: ☠
Ingredients: Copper sulphate pentahydrate ($CuSO_4*5H_2O$); water (H_2O)
Core process: Crystallisation

✓ In nature ✓ In lab ✓ At home ✓ Toxic

Copper sulphate creates brilliant blue crystals when its hydrated form is dissolved in hot water. Copper sulphate is a type of salt, most commonly encountered as a powder – copper sulphate pentahydrate (CuSO4*5H2O). This is a way of expressing five water molecules attached to the copper sulphate molecule; it is hydrated. For blue crystals to form, copper sulphate pentahydrate is added to hot water until no more can dissolve. This is referred to as a saturated solution, and a hotter solution can dissolve more copper sulphate than a colder one.

When the solution cools, some of the copper sulphate can no longer exist in a dissolved state, so the molecules gather in an organised repeating pattern, forming crystals. This is a physical change – the material is altering its structure, not its makeup. Suspending a nylon wire in the solution creates a surface for the crystals to latch on to, encouraging growth. Eventually the water evaporates, but copper sulphate can't so it's forced into an ever-smaller space. The molecules of copper sulphate continue crystallising until no water is left.

Top 5 deadliest reactions

1 Hydrochloric acid
and most things...
Hydrochloric acid (HCl) is an extremely strong acid. This acid reacts with most things – especially bases – and can corrode metal, cause chemical burns and even release flammable hydrogen.

2 Acid rain
When sulphur dioxide is released into the air, it rises up and reacts with hydrogen peroxide, which is found in some clouds. Sulphuric acid – a product of the reaction – falls back down to Earth and can have devastating effects on flora, fauna and buildings/statues, etc.

3 Nitroglycerine and heat
Nitroglycerin is one of the most explosive substances there is. The oily liquid is so sensitive that the slightest jolt or increase in heat can trigger a massive explosion.

4 Bleach and ammonia
When ammonia and bleach are mixed, the bleach decomposes to form hydrochloric acid. Ammonia and chlorine gas react in order to form a deadly vapour: chloramine.

5 Mustard gas
The volatile combination of sulphur dichloride and ethylene reacts to form a cyclic sulphonium ion. This reacts with parts of DNA to prevent cells from replicating, leading to tissue necrosis.

The firestarter

Burning hydrogen

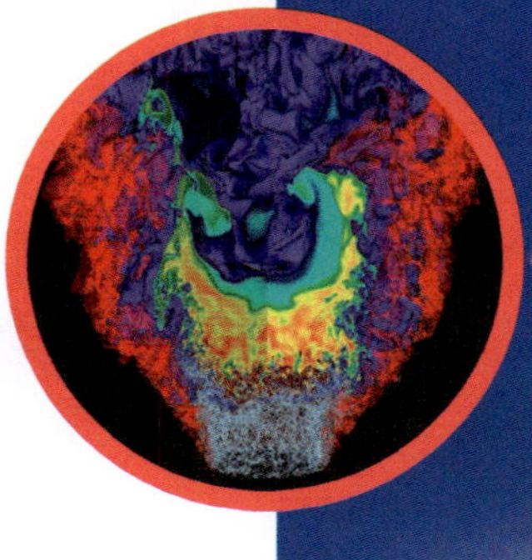

Deadliness: ☠☠☠
Ingredients: Hydrogen (H_2); oxygen (O_2)
Core process: Redox

✖ In nature ✓ In lab ✖ At home ✖ Toxic

Hydrogen (H_2) is the lightest, most abundant element in the universe, yet it's also one of the most flammable. Hydrogen is quick to burn in the presence of oxygen (O_2) and can be very explosive. Used as the primary fuel for combustion when launching space shuttles, this is seriously powerful stuff. When hydrogen burns, large quantities of heat and light are given off. The light emitted from a pure hydrogen and oxygen reaction is mainly ultraviolet, making the flame almost invisible – however, in reality, there are often other materials present, creating a visible flame. Water is the waste product of hydrogen combustion, since oxygen and hydrogen are the two ingredients in water. Combustion of liquid hydrogen and oxygen is used to launch rockets – hence it is water vapour, not smoke, that you see coming out of the exhaust during the takeoff.

Scientists are now working on using hydrogen combustion to power cars and other machines. The difficulty is the large amount of initial energy needed to get the reaction going. It requires far more energy to get started than, say, traditional fossil fuels.

Hydrogen is rarely found on Earth in its pure form, because it prefers to join with other elements – and of course a great deal exists as water.

The bomb decoy

Magnesium and Teflon

Deadliness: ☠☠☠
Ingredients: Magnesium (Mg); Teflon ((C_2F_4)n)
Core process: Redox

✖ In nature ✓ In lab ✖ At home ✖ Toxic

Magnesium (Mg) is a highly reactive element. which burns at a staggering 3,100 degrees Celsius, giving off an intense white light. In addition to visible light, magnesium emits infrared (IR) when burned, making it perfect for use in military countermeasures such as decoy flares. Like all things, magnesium needs to be in the presence of an oxidiser when it burns – a material which takes electrons from the fuel allowing the reactic to occur. Flares are made of Teflon ((C_2F_4)n) and magnesium, and it's the fluorine in Teflon that oxidises magnesium. Fluorine is a stronger oxidiser than oxyger as it wants to accept electrons more than oxygen, allowing for a higher temperature of combustion. Heat-seeking missiles lock on to infrared light given off by engines in aircraft, but magnesium decoy flares throw out far more IR light than aeroplane engines, effectively confusing the missiles' heat-seeking guidance systems and hopefully deterring the weapon from its target.

A magnesium fire cannot be extinguished with water, since the magnesium reacts with water to produce hydrogen gas – which if anything will only intensify the fire. Instead, dry sand is generally used to stop the reaction. Other uses of magnesium have been as an illumination source in flash photography and in firework

The solidifier

Sodium acetate supersaturation

Deadliness: ☠
Ingredients: Sodium acetate (NaC_2H3O_2); water (H_20)
Core process: Crystallisation

✗ In nature ✗ In lab ✓ At home* ✗ Toxic *(Heat is given off but not enough to cause burns)

odium acetate ($NaC_2H_3O_2$) heated in water then cooled has the unusual property of crystallising into a solid when it is disturbed. It can be poured out of a beaker as a liquid and, upon hitting a surface, becomes a solid that is hot to touch – hence its other name, hot ice. Sodium acetate is a salt which dissolves in water. Heating – to around 100 degrees Celsius – then cooling a mixture of the two enables more sodium acetate to dissolve to form a supersaturated solution. The solution exists in a metastable state, analogous to a ball perched at the top of a hill, where the slightest of nudges will make it roll down.

The trigger can be pouring the solution out of the container, or adding a seed crystal, causing the dissolved sodium acetate to come out of the solution and return to a solid. In our analogy this is like the ball rolling down the hill until it reaches flat ground and a lower energy state.

Along the way, the solid sodium acetate absorbs three molecules of water, becoming sodium acetate trihydrate ($NaC_2H_3O_2{*}_3H_2O$). These water molecules are not chemically bonded to the sodium acetate, representing a physical change. The process is exothermic (ie it releases heat) and, as a result, it's often used in hand warmers.

The fractal maker

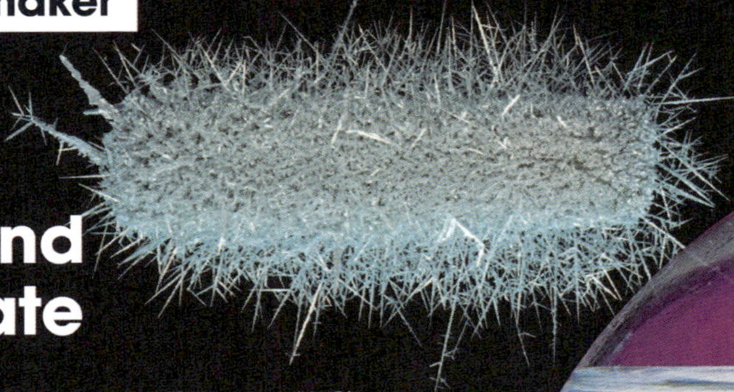

Copper and silver nitrate

Deadliness: ☠
Ingredients: Silver nitrate ($AgNO_3$); copper (Cu); water (H_2O)
Core process: Single replacement

✗ In nature ✓ In lab ✗ At home ✓

Mixing silver nitrate and copper is one of the most famous chemistry experiments, with it starring in many school science lessons around the globe. The experiment involves introducing copper – typically a copper wire – to a silver nitrate/water solution and suspending it there for a couple of hours.

The combining of both triggers a single replacement reaction, where copper is changed from its elemental form (Cu) to its blue aqueous ion form (Cu2+ (aq)), while the silver ions (Ag+ (aq)) in the silver nitrate solution will be changed into their elemental metallic form (Ag) and deposited onto the wire. These silver deposits continue to grow off the copper in a series of fractal-like crystals until all reactable copper in the solution is exhausted, leaving the end products of silver and copper nitrate.

The reason this replacement reaction occurs is that the atoms in the copper are oxidised when introduced to the silver nitrate solution, losing electrons and forming copper ions, while the silver ions in the nitrate solution are reduced (ie they gain electrons) into elemental silver.

What's really cool is that once the silver crystals have grown they can be removed from the copper, dried off and then displayed as funky pieces of fractal art.

The witch's potion

Potassium and water

Deadliness: ☠☠☠ **Ingredients:** Potassium (K); water (H_2O) **Core process:** Redox

✗ In nature ✓ In lab ✗ At home ✗ Toxic

Put a lump of potassium in a dish of water and it will give off a pinkish light, get very hot and skim across the surface at speed. A favourite experiment of many science classrooms, potassium is a highly reactive metal that reacts violently in the presence of oxygen and water. It forms potassium hydroxide (KOH) and hydrogen gas (H2). Potassium atoms have 19 electrons – one of which is alone in an outer shell. This makes potassium very keen to lose an electron so it has a complete outer shell and is more stable. When the reaction begins, enough heat is given off to ignite the hydrogen gas, which then reacts with oxygen to produce water. Potassium is so reactive that it must be stored in kerosene, so as not to come into contact with water vapour in the air. Even oxygen in the air is enough to cause potassium to spontaneously combust! Because potassium is so reactive, it's not found in its elemental form, but is common as a compound.

The screaming Jelly Baby

Potassium chlorate and most things (in this case Jelly Babies)

Deadliness: ☠☠☠ **Ingredients:** Potassium chlorate (KClO3); Jelly Babies (glucose syrup, sugar, water, gelatine and flavourings)
Core process: Redox

✗ In nature ✓ In lab ✗ At home ✓ Toxic

Watching a Jelly Baby meet its demise at the hands of potassium chlorate is a spectacular affair. There's an abundance of energy inside Jelly Babies stored as sugar, released in intense flames and a piercing scream when potassium chlorate is added to the mix.

Potassium chlorate is a powerful oxidiser, taking its form as a white powder and commonly used in fireworks and explosives. The 'ate' part of chlorate describes the oxygen atoms attached to the chlorine atom, and the chemical formula is KClO3. Chlorate-based oxides are more efficient oxidisers than those in gunpowder and potassium chlorate needs to be handled very carefully due to its unpredictable ability to spontaneously ignite.

The reaction happens when a small amount of potassium chlorate is placed in a test tube and heated until it becomes a clear liquid. Needless to say, safety screens and goggles are a must. The Jelly Baby is placed with tongs into the tube and instantly produces lively flames, intense screaming and plenty of smoke. The reaction can last up to 20 seconds and gives off noxious fumes, so ventilation is also needed.

Ferrofluid

This experiment combines physics and chemistry to produce an awesome effect. Ferromagnetic fluid is a liquid that undergoes a radical change when introduced to a magnetic field, turning from a puddle into a spiked dome. The fluid does this due to its composition, which is a mix of nanoscale ferromagnetic particles (like iron) and a carrier fluid. The particles are coated with a surfactant – a compound that lowers a liquid's surface tension – ensuring an even distribution of particles. When a magnetic field is introduced – usually a strong magnet positioned beneath it – the particles realign to the magnetic field lines. Contained as they are, the particles cause the liquid to act like a solid.

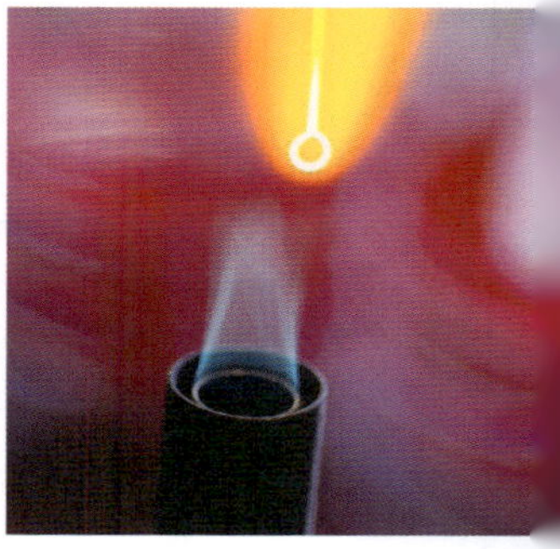

Flame test

The flame test is one of the simplest yet coolest experiments in the lab. By introducing certain elements – generally metals – to a Bunsen burner, you can determine their composition by analysing the emission spectrum. This works as the heat excites the material's ions, so they emit visible light. For example, if you have a chunk of unknown metal, by introducing it to a calibrated burner (one that is not contaminated) and evaluating the colour(s) of the flame, you can determine what the substance is made of. Copper (Cu) emits a blue-green flame, lithium (Li) a bright red one, while the image above shows the orange/crimson flame generated by strontium (Sr).

The barking dog

·arbon disulphide and nitrous oxide

·eadliness: ☠☠☠
·ngredients: Carbon disulphide (CS_2); nitrous oxide (N_2O)
·ore process: Redox

✓ In nature (the reactants) ✓ In lab ✗ At home ✓ Toxic

·he barking dog reaction is a ·onsequence of igniting ·arbon disulphide (CS_2) mixed ·ith nitrous oxide (N_2O) – the ·atter is better known as ·aughing gas. The reaction ·enerates a bright flash of ·lueish-purple light and heat, ·nd, more bizarrely, a sound ·ke a dog barking.

Nitrous oxide gas is the ·ource of oxygen – ie the ·xidiser – needed to burn the ·olourless liquid fuel, carbon ·isulphide. When the reaction ·akes place in a confined ·pace – such as a long tube some energy is converted to ·orm the rapid but loud barking ·oise, due to a fluctuation of ·ressure. This is an example of · reaction which makes elements from compounds: in this case a yellow coating of sulphur and nitrogen gas are the elements left in its wake.

Carbon disulphide is found in nature as a product of the metabolic processes in plants, and also volcanic eruptions. Nitrous oxide also forms naturally from some species of bacteria, plus through industry and agriculture, and it depletes ozone in the stratosphere.

Used in the distant past as a method of flash photography, the flash it produces is so bright that many people in the photographs would often appear startled. The pervasive smell that sulphur compounds are capable of probably didn't make it that popular either.

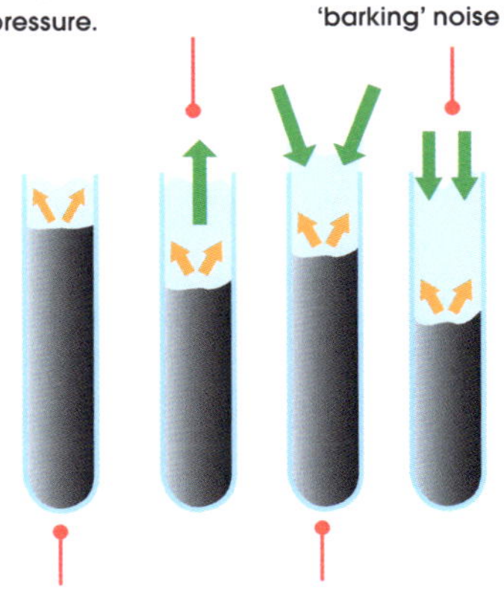

1 Energy release Nitrous oxide reacts with the carbon disulphide releasing energy as heat, expanding the gases.

2 Expansion As the gases expand, those near the top are forced out of the test tube due to pressure.

3 Differential The expelled gases lead to a pressure drop within the tube, creating a vacuum-like effect.

4 Bark As the gases rush back into the tube to balance the pressure, a repeated 'barking' noise is made.

The candy rocket

Potassium nitrate and sugar

Deadliness: ☠☠
Ingredients: Potassium nitrate (KNO_3); sucrose ($C_{12}H_{22}O_{11}$)
Core process: Redox

✗ In nature ✓ In lab ✗ At home ✓ Toxic

There's something strangely satisfying about witnessing the volatile display of smoke, colour and fire given off by the mixture of potassium nitrate, sugar and heat. The amount of fire varies, but there is always an abundance of smoke. You have most likely seen this reaction at a firework display, or from the smoke stunt planes deploy where coloured dyes are often added for effect.

Potassium nitrate (KNO3)– aka saltpetre – is an essential in any pyrotechnics cookbook; it's one of the main ingredients in gunpowder, for example. Potassium nitrate works as an oxidiser, giving off oxygen and promoting the burning of fuel.

As seen in the 'Screaming Jelly Baby', sugar is an extremely effective fuel; it contains energy that 'burns' in our bodies and converts to useful energy we use to perform any physical activity. When heat is applied to saltpetre and sugar, the saltpetre loses an oxygen atom – transitioning from KNO3 to KNO2 and oxidising the sugar. The sugar burns, releasing smoke which rapidly expands and can generate enough thrust to lift a small rocket.

Interestingly there was a programme called Sugar Shot to Space that aimed – as you would probably guess – to launch a rocket powered by sugar propellant alone beyond Earth's atmosphere.

Top 5 useful reactions

1 Photosynthesis
The chemical process of turning sunlight into energy is vital to life, enabling plants to grow and release their waste product: oxygen. As well as regulating the levels of oxygen in the atmosphere, photosynthesis is the source of energy for most organisms.

2 Baking bread
In times when other food sources were scarce, the discovery of heating flour mixed with water combined with yeast fermentation was nothing short of a breakthrough. Yeast ferments sugars and carbohydrates, which gives off CO2.

3 Extraction of metals
One of the world's largest industries, extracting metals from their ores using chemical reactions is hugely important, as most metals are mixed with impurities in their natural state.

4 Galvanising steel
From lampposts to buildings, steel is everywhere around us and protecting these structures from rust is vital. Dipping steel in a bath of zinc causes a chemical reaction that adheres a coating of zinc to the steel. This coating protects it from water vapour.

5 DNA production
The essential 'code' for all known life, DNA is a molecule with our genetic instructions in the form of nucleotides – a set of long polymers made of sugars and phosphates. Chemical reactions allow the DNA to form, replicate and interact with proteins to make us who we are.

Explosive action

Humans have created many different types of explosives over the past 1,000 years, deploying them in warfare, industry and society

Explosions are characterised by three factors: a rapid increase in volume, the generation of high temperatures and an extreme release of energy. In addition to this, every explosion generates a shock wave that varies depending on the explosive substance's detonation velocity.

The four main properties of explosions are force, velocity, heat and fragmentation. The force of any explosion is the physical impact it has on its immediate environment, something often increased through a warhead's shaped charge. An explosion's force when magnified in a shaped charge can gauge impact craters, tear through the armour of military vehicles and puncture buildings.

An explosive's detonation velocity is the rapidity of the expansion of its volume (gases mainly) post detonation. Typically, the greater the detonation velocity the higher up the relative effectiveness (RE) table the explosive is placed, with a direct correlation between velocity, pressure and generated heat. So, generally, the greater the velocity the greater the heat and damage produced.

Heat, generated by and also the cause of an explosion's high pressures, is its ability to liberate its exothermic material's elements at a certain temperature. Coal, for example, is very good at liberating its material's elements at a high temperature in an exothermic reaction, albeit at such a slow rate that it cannot be used as an explosive (why we use it for fires). Nitroglycerin (the chemical base of explosives such as dynamite) produces five times less heat as coal but has the necessary rapidity in liberation for an explosion to be caused.

Finally, almost all explosions are subject to fragmentation – the accumulation and projection of particles as a result of its detonation. Fragmentation is proportional to the explosion's generated shock waves and detonation velocity, with supersonic shock waves (those travelling faster then the speed of sound) producing considerably more fragmentation than subsonic varieties. Fragmentation increases an explosion's potenc ten-fold as low-angle high-velocity shards can shred human tissue, render vehicle armour and level structures.

The history of explosives

Chart the rise of manufactured explosives over the centuries

800

Ancient Chinese: The ancient Chinese are credited with the creation of early explosives – which were crude black powders – used in rockets and fireworks.

1300

European gunpowder: By the 1300s, the knowledge of the East had reached Europe and a new era of explosive warfare ensued, relegating suits of medieval armour to museums.

1846

Nitroglycerine: More powerful than black powder, nitroglycerine paved the way for the creation of dynamite and other explosives.

Why are explosions deadly?

Capable of breaking the physical threshold of human tissue, explosions claim many lives

This Schlieren photograph shows a reconstruction of a bomb exploding beneath the passenger seat of an aircraft. Notice how the blast generates circular shock waves of heat and light energy that radiate vertically within the cabin

Explosions are deadly for a variety of reasons. The basis of any explosion is the rapid burning and decomposition of a material, a process that creates a large amount of heat, pressure and gas in a short amount of time. All of these factors can have an effect on humans, as their physical thresholds to withstand the various forces and chemicals in play is easily broken. For example, the heat generated by an explosion of HMX (a nitroamine high explosive) reaches thousands of degrees Celsius in just a matter of seconds. This is something the skin of a human – which would swiftly be subjected to third-degree burns – could not withstand, with the skin tissues and structures destroyed.

The force and velocity of the detonation is equally as deadly. The shock wave created by an explosion carries tremendous amounts of energy and can easily project objects of such light weight hundreds of feet. Fragmentation – the shattering of explosive casing and material in an explosion's immediate vicinity – in supersonic explosions can also cause severing and puncturing to human tissue. In addition, the force of any explosion is often increased if it is a consequence of the detonation of an explosive device. This is because many warheads, mortars, rockets, shells and bombs have shaped charges, as demonstrated in the example of C-4, a type of plastic explosive based on cyclotrimethylene-trinitramine (RDX). C-4 is deadly as it can be moulded into various shapes to increase its destructive potential.

Did you know?
Despite octanitrocubane's massive explosive power, it is shock-insensitive and not easily detonated

Gunpowder

Since its invention by the ancient Chinese, gunpowder has been one of the most widely-used explosives on Earth

Gunpowder, unlike other modern explosives, is classified as a low explosive – one that undertakes subsonic combustion – as it deflagrates slowly. For this reason, despite its creation over 1,000 years ago, it was still used as the explosive mechanism in guns up until the 20th century, providing enough energy to propel a bullet out of a muzzle, but not enough to break the weapon and injure the user (of course, backfires and jams lead to bad injuries).

Modern rifle cartridges – the combination of a bullet, case, powder and primer – such as the one fired in the accompanying image, use smokeless powders (they are not actually smokeless, however, just considerably less so than conventional black powder), as their combustion efficiency is greatly increased, enabling rounds to be fired with less explosive. Further, smokeless powder does not leave heavy fouling (unwanted residue materials) within the weapon, allowing for semi and full automatic rifles to operate cleanly and efficiently.

This image shows the shock waves from the passage of the supersonic bullet and the muzzle blast, as well as a white cloud of propellant gas

1863

rinitrotoluene: Trinitrotoluene or TNT for short, was created by German emist Joseph Wilbrand in 1863. It is ow used as the modern benchmark for all other explosives.

1950

C-4: A common type of plastic explosive due to its easy handling, C-4 is based on the explosive RDX and provides massive, yet refined, damage.

2000

Octanitrocubane: With performance roughly 25 per cent greater than HMX, this is currently considered the most powerful non-nuclear explosive on the planet.

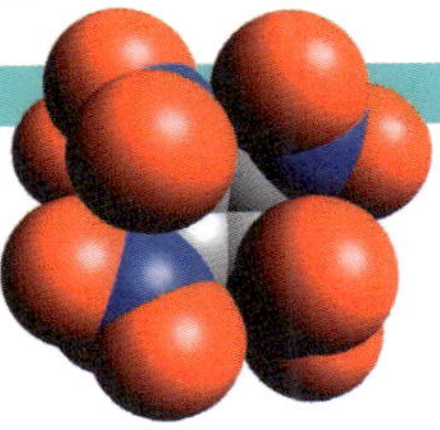

You were only supposed to blow the doors off!

An Italian Job-style visual guide to the six most powerful non-nuclear explosives on the planet

Trinitrotoluene

This was the first real breakthrough for explosives post-gunpowder. Trinitrotoluene is one of the most commonly used explosives on Earth. It was formed in 1863 and has an RE factor of 1.00.

Molecular formula:

$C_6H_2(NO_2)_3CH_3$

1.5/5

Small explosion. Detonation velocity of 6,900m/s.

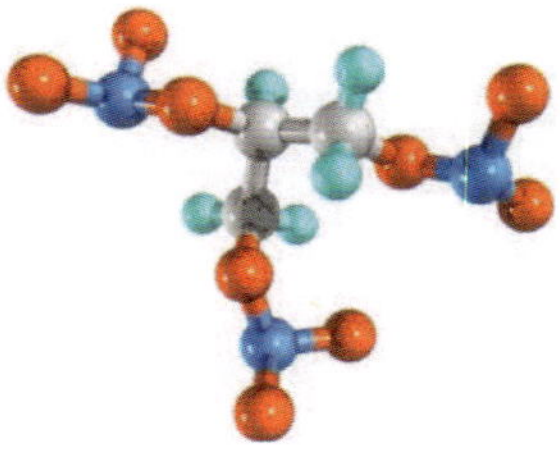

Dynamite

Based on nitroglycerin, dynamite can boast a whopping 60 per cent greater energy density than TNT. It was created in 1867 by the renowned German chemist Alfred Nobel and has an RE factor of 1.50.

Molecular formula:

$C_3H_5N_3O_9$

2/5

Modest explosion. Detonation velocity of 7,700m/s.

C-4

Based on the explosive RDX, C-4 is the most famous plastic explosive used today. It is easily moulded into any shape (excellent for shaped charges) and has a high detonation velocity. It has an RE factor of 1.34.

Molecular formula:

$C_3H_6N_6O_6$

2.5/5

Large explosion. Detonation velocity of 8,040m/s.

Semtex

Another brand of plastic explosives used mainly in commercial and industrial applications, Semtex is a mi of RDX and PETN. It is brick-orange in colour and has been given an RE facto of 1.66.

Molecular formula:

$C_5H_8N_4O_{12}$

3/5

Great explosion. Detonation velocity of 8,420m/s.

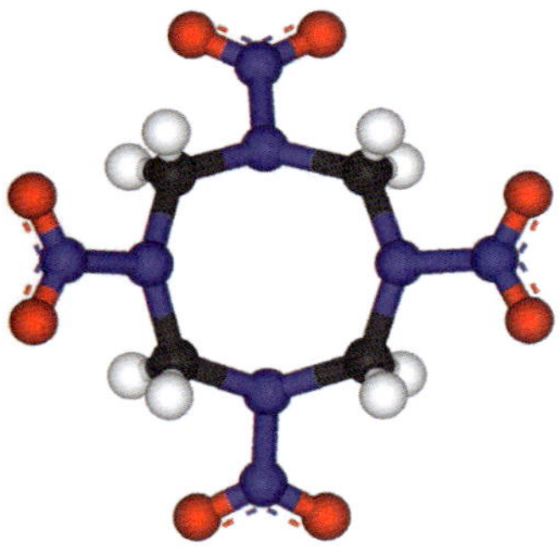

HMX

Related to RDX but much more explosive, HMX is a nitroamine high explosive. Difficult to manufacture, it is only used for specialist military purposes, such as a nuclear bomb detonator. It has an RE factor of 1.70.

Molecular formula:

$C_4H_8N_8O_8$

4/5

assive explosion
etonation
elocity of 9,100 m/s.

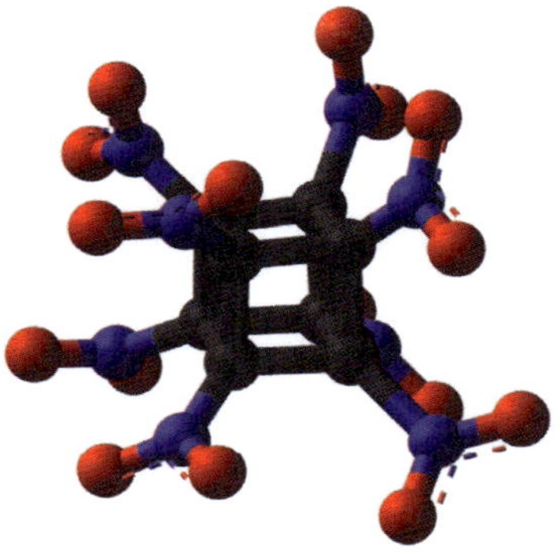

Octanitrocubane

The most explosive chemical structure on Earth, Octanitrocubane has 25 per cent greater performance than HMX. It was synthesised at the University of Chicago in 1999 and has an RE factor of 2.70.

Molecular formula:

$C_8(NO_2)_8$

5/5

Epic explosion. Detonation velocity of 10,100 m/s.

Explosions in nature

Explosions are not always made, with many natural examples visible over the Earth and in space

Organic: Many explosions are caused by plants and trees when subjected to hot and volatile conditions. For example, the Eucalyptus tree – which is predominantly found in Australia, New Guinea and Indonesia – is often prone to spontaneous combustion in periods of sustained heat and dry conditions. This is caused by the tree's high oil content (ie Eucalyptus oil) that rises above the bush as vapour when heated sufficiently, creating a characteristic haze in dense forested areas. This oil, in liquid and vapour forms, is incredibly flammable and if environmental pressure and heat reach critical levels, is prone to combust. This factor is severely exacerbated when a forest fire passes into Eucalyptus-heavy areas, as the resulting explosions and general flammability of the tree's bark carries and generates fire quickly.

Volcanic: Maybe the most obvious example of a natural explosion, volcanoes cause explosions when magma, which has risen to the Earth's surface, releases large amounts of accumulated dissolved gases into the atmosphere. This reduction in pressure above ground causes the gas to be separated from the magma and rapidly increase in volume, sending massive clouds of volcanic dust and particles spiralling into the atmosphere. Further, if a volcano's magma flow reaches a water source, such as the ocean, then further steam-based explosions may be caused when the high-temperature flow meets the cold water, rupturing sections of coastline.

Stellar: The largest natural explosion currently known in the universe is that of a supernova, the resulting explosion that follows a dying star. Supernova explosions are ridiculously powerful and send the remnants of the star's material screaming out through space at a velocity up to 30,000km/s – that is ten per cent the speed of light – in a vast shock wave. Supernovas can be caused in a variety of different ways, however the most common is when the core of an ageing star ceases to generate nuclear fusion, collapsing in on itself under the effects of gravity and expelling its outer layers as it releases its gravitational potential energy. During its short explosion (due to its size this would last several weeks or months) the amount of radiation of a supernova can total the entire amount it has ever radiated over its whole life span.

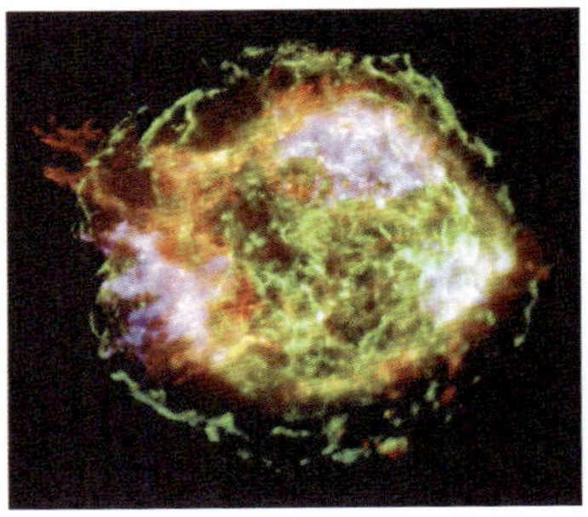

Matches

What is the substance that brings us portable flames?

A match consists of two main parts: a head and a tinder substance, the latter which also acts as a handle. The part of the match that starts combustion is the head, a coating of chemicals that initiates a flame once enacted upon by an external force and energy, a process that on modern matches involves striking them against a rough surface. The igniting agent on the majority of match heads is phosphorus – due to its high reactivity – and this is partnered with one or multiple oxidising agents such as potassium chlorate, which help supply oxygen to the tinder substance and stabilise the lit flame.

In addition to an igniting agent and oxidiser, match heads are also coated with binders – animal glue, starches and gums, for example – which help bind the other chemicals, as well as various inert materials such as diatomaceous earth that provide bulk and control the speed of reaction. Further, in the case of safety matches – matches that can only be lit on a special, pre-prepared surface – additional reactive chemicals are applied to the striking surface. Then, once a flame has been ignited and stabilised, it continues to burn until the tinder is decimated.

Head
The majority of match heads consist of a mixture of phosphorus and oxidising agents.

Tinder
The tinder also acts as the handle, commonly made of wood.

Safety matches
These can only be lit on a pre-prepared surface that has been coated with reactive chemicals.

How does paint dry?

Pivotal to its application, the drying mechanism of paint tells us much about its formulation

The majority of paints dry through evaporation, a process that allows its pigmentation to be set onto a chosen surface purely by being exposed to the surrounding atmosphere. However, the drying process can vary and is often complicated and altered between different paint types.

Paints usually contain three key components: pigments, binders and additives. Pigments are dry, insoluble powders that by wavelength-selective absorption (ie they only reflect certain wavelengths) change the colour of reflected or transmitted light, giving paint its colourisation. They can be ascertained either naturally or produced synthetically.

The pigments of paint are given their paint structure by binders, synthetic or natural resins such as acrylics, polyesters or oils that impart adhesion and influence durability and flexibility. Crucially, though, binders can also play a role in how paint dries, curing it as well as supplying it with adhesion. It is important to note that curing is a different process to drying – which as aforementioned, is caused by evaporation of a solvent – with cure adhesion attained by polymerisation (molecules bonding together in a chemical reaction). Binders are arguably the key component of paint, as without them it would never stick to a surface long enough to dry.

Finally, paint additives help to conjoin the other components and aid application, structure and drying. Certain additives are often used as catalysts for polymerisation, while others are included to prevent the clumping of paint or in order to alter its viscosity.

The process
Paint dries primarily by utilising a solvent. Watercolours, for example, set when their water molecules evaporate in the atmosphere.

Pigmentation
Pigment provides colour by absorbing wavelengths and reflecting others.

How smoke detectors work

They may annoy us when toast burns, but these ear-piercing devices save lives

There are two main types of smoke detector: optical and ionisation. Optical detectors contain an infrared light beam pointing toward a photocell, which generates electricity when light alls on it (like on solar panels). When there is no moke, the light reaches the photocell unobstructed. his is registered by the internal circuitry so the larm is not triggered. When there is a fire, smoke nters the detector and blocks the beam of light, o the photocell can no longer produce an electric urrent. This change is picked up by the circuitry, riggering the alarm and alerting people to danger.

Ionisation detectors contain a small sample of radioactive substance, typically americium. This onstantly emits alpha particles (positively-charged elium nuclei), which pass between two charged etal plates called electrodes. The alpha particles ollide with air molecules and split them into ositive ions and negative electrons. These charged articles are attracted to opposite electrodes, ausing a current to flow. Smoke particles can ttach to ions and neutralise them, so they are no nger attracted to the electrodes. A sensor detects e drop in current and the alarm is triggered.

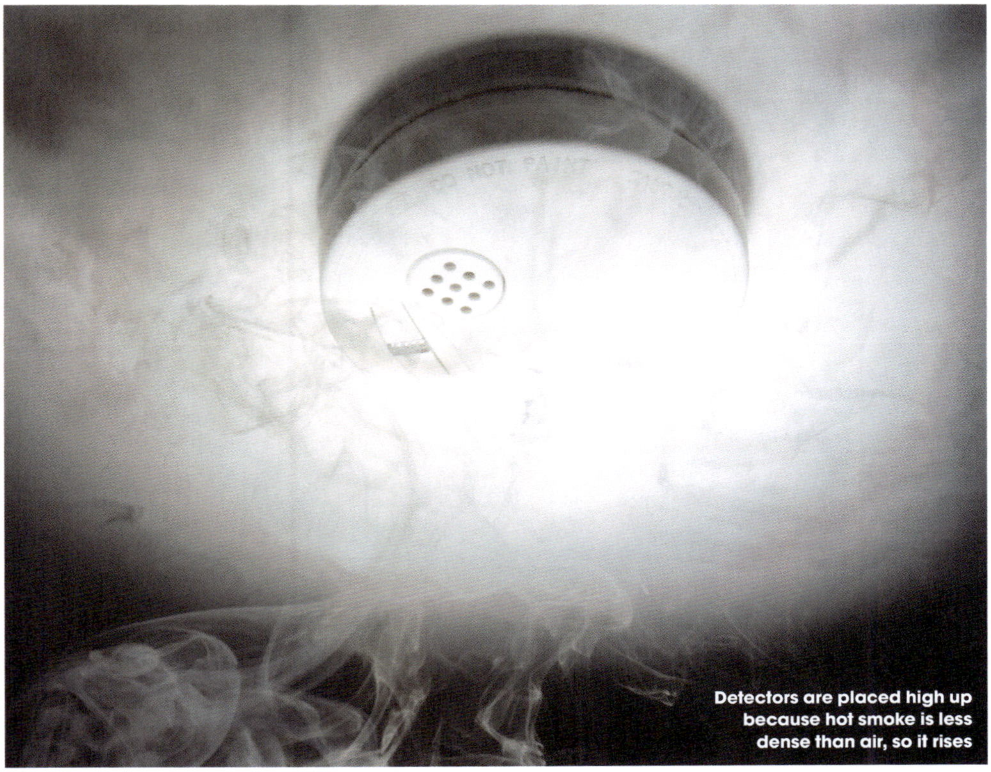
Detectors are placed high up because hot smoke is less dense than air, so it rises

As wood burns, hydrocarbons evaporate into the air, either bursting into flame or floating up as smoke

What is smoke?

Find out what happens when a fire burns without enough oxygen

Wood is made from a combination of water, hydrocarbons, and minerals. As the temperature rises above 149 degrees Celsius, the hydrocarbons start to vaporise, floating up into the air. They combine with oxygen, burning and releasing energy that is visible as a hot flame. As they burn, they release carbon dioxide and water vapour, both colourless gases that you cannot see.

Sometimes there is not enough oxygen for these evaporated components to burn, and instead, they continue to float upwards, rising in the hot air. The evaporated oils and tars clump together as they rise, forming fine particles known as smoke. The particles move about randomly, and as they collide with the invisible particles that make up the air, they change direction, producing the ever-changing swirling patterns of smoke.

After all of these volatile hydrocarbons have evaporated, all that is left is charcoal, nearly pure carbon. The charcoal does not evaporate, so as it burns there is no flame, just glowing embers. Once the charcoal is gone, all that remains is ash, the minerals like magnesium and potassium that do not burn at all.

How to tame fire

From the chemistry of candles to flames in space, discover the science behind this hot topic

Burning underwater

Some materials burn so furiously that not even water can put them out. Used in metal refineries and even bomb production, thermite is a human-made combination of powdered aluminium and a metal oxide, often iron. Once ignited, this marriage of metals can burn at an alarmingly high temperature, around 2,400 degrees Celsius, spewing molten chunks of iron on the ground.

Back in 2015, a popular YouTube channel called 'TheBackyardScientist' wanted to see if the reaction could be halted when dipped in water. Sure enough, not even water could stop the thermite reaction from burning while submerged until the fire ran out of fuel.

Ever since our ancestors rubbed two sticks together to produce embers, humans have evolved a fascination with fire. In a relationship which has defined our modern-day selves, fire has transformed the way humans eat, stay warm and build global civilisations. It was described in ancient Greece as one of Earth's 'natural elements' alongside water, air and earth. But what exactly is fire?

Fire is an interaction of particles that creates a hot, glowing plasma. Fundamentally, what we see when we strike a match is a series of chemical reactions that emit heat, light, gas and water as byproducts. Every flicker and flare of a dancing flame is created by atoms releasing energy at different stages of being heated.

The recipe for fire is relatively simple. Add a dash of oxygen, a drizzle of fuel and a heap of heat, and hey presto, you've got fire. While all three are required to create fire, the fuel source can be almost anything. The two most flammable elements on Earth are carbon and hydrogen, so fuels that are heavily comprised of these elements are a fire's favourite food. This is because carbon and hydrogen like to combine with oxygen in the air. With the initial release of heat energy such as a spark, a fuel source like wood needs to reach its own ignition temperature in order to burn. Once heated up, volatile gases such as carbon and hydrogen released from the fuel begin a chain reaction of oxidation called combustion. During this rapid oxidation, the chemical bonds that make up the fuel source are broken apart and reformed into the byproducts of the fire, carbon dioxide and water. In this process of ripping molecules

apart, energy is released in the form of heat and incandescent light.

However, not all fires are created equal when it comes to how hot they burn. A fire's temperature can vary wildly depending on the chemical composition of the fuel and the amount of oxygen available. For example, a wood fire may burn at around 1,100 degrees Celsius, whereas ethanol can burn at around 1,920 degrees Celsius. This is predominantly due to the higher proportion of carbon and hydrogen in ethanol compared to wood. Another important factor that can determine whether or not something catches fire is its size. Thicker and denser fuel sources, such as a tree trunk, understandably take a great deal longer to reach a temperature where they might ignite than a smaller fuel, such as kindling.

Fire is indiscriminate in its search for fuel, and this can lead to devastation. By the very nature of this chemical reaction, until a fuel source is completely depleted or a critical ingredient of the fire recipe is removed, flames will continue to rage, destroying everything in their path. One of the biggest examples of the untameable nature of fire are the 2020 bushfires in Australia. Raging

Wildfires can be speedy, travelling up to around ten kilometres per hour in forests and 23 kilometres per hour in grasslands

Types of extinguishers

Using the wrong one can make things a lot worse

Wet chemical

- Solid material (wood and paper)
- Cooking oils

Due to the volatile reactions between water and hot oil an alternative wet chemical extinguisher is used to tackle these fires. These extinguishers disperse a mist of potassium salts that react with the burning oil to create a soapy film, cutting off oxygen supply.

Water

- Solid material (wood and paper)

Only suitable for solid material fires, such as wood and fabrics, water removes heat from the burning fire, cooling it down and preventing it from igniting again. Water extinguishers should never be used on electrical or burning fat or oil fires.

Carbon dioxide

- Flammable liquids
- Electrical

In fires where electricity is at play, using liquid extinguishers can spread the fire and make matters decidedly worse. In these cases, carbon dioxide gas extinguishers are employed to cut off air supply to the fire.

Dry powder

- Solid material (wood and paper)
- Flammable liquids
- Gaseous fires
- Electrical

The most versatile type of extinguisher – a dry powder extinguisher sprays a mixture of sodium or potassium bicarbonate or monoammonium phosphate in order to coat and smother a variety of fire types.

Foam

- Solid material (wood and paper)
- Flammable liquids

This type of extinguishe works by creating a sea over flaming liquid or burning wood. By doing this, it prevents any flammable vapour from igniting surrounding oxygen, thus starving th fire into submission.

Hot rainbow

Changing the colour of fire might be an experiment you remember from school. When heated, different elements emit an array of colours as light energy, a result of the elements' electrons 'jumping' from a normal to an excited state. As they fall back down to a normal state, they emit wavelengths of differing colours. For example, copper burns with a blue-green flame. Larger atoms such as strontium require less energy to excite their electrons, and so during their energy comedown emit a colour with a longer wavelength – red. However, smaller atoms such as sodium demand more heat energy to get excited, therefore having more energy to expel as colours with shorter wavelengths.

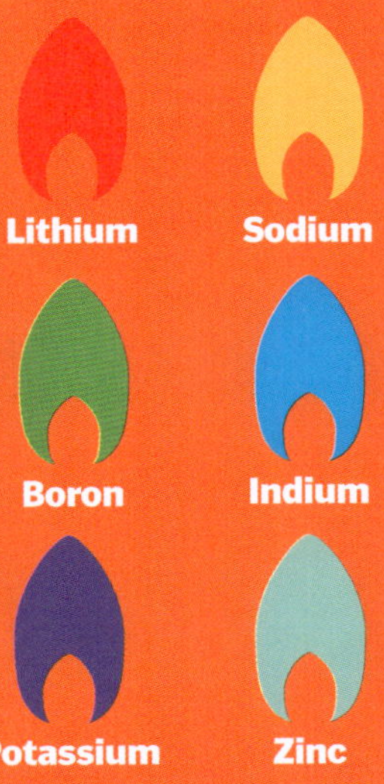

Candle combustion

Fire isn't just destructive. It can light our homes or be used in celebration

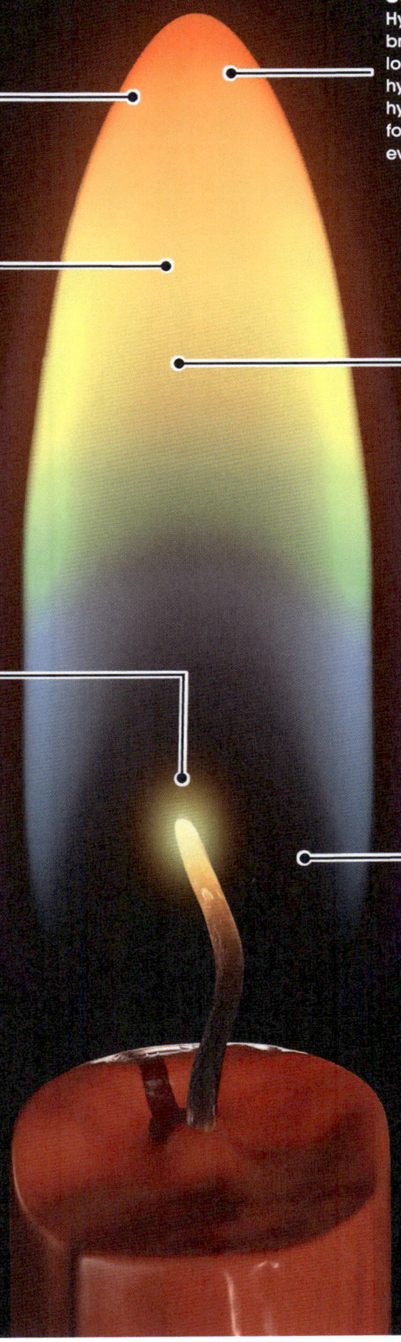

Light
The majority of light produced by a candle is emitted from the surface of the flame, rather than the hottest region. Visible light is emitted largely by the burning of incandescent soot molecules moving up the flame.

Soot
Hydrogen and carbon that are broken apart are reformed as loop-like polycyclic aromatic hydrocarbons. These hydrocarbons clump together to form soot, rising up the flame to eventually become smoke.

Oxygen
Drawn in from the air, oxygen combines with free carbon atoms and is released as carbon dioxide.

Water
In the same way that oxygen combines with a candle's carbon, oxygen also comes together with hydrogen in order to produce water as another by-product.

Wax
As a candle's fuel source, melting wax is pulled up the wick when set alight, where its hydrocarbon molecules that meet the flame are vaporised and break apart.

Fire extinguishers are designed to smother fires by removing one of three vital ingredients needed for their creation

Heat
During the reactions between oxygen, carbon and hydrogen, energy is released as visible infrared and heat, with the hottest portion of the flame located at the base.

Inside a fire extinguisher

How water extinguishers use pressure to put out fires

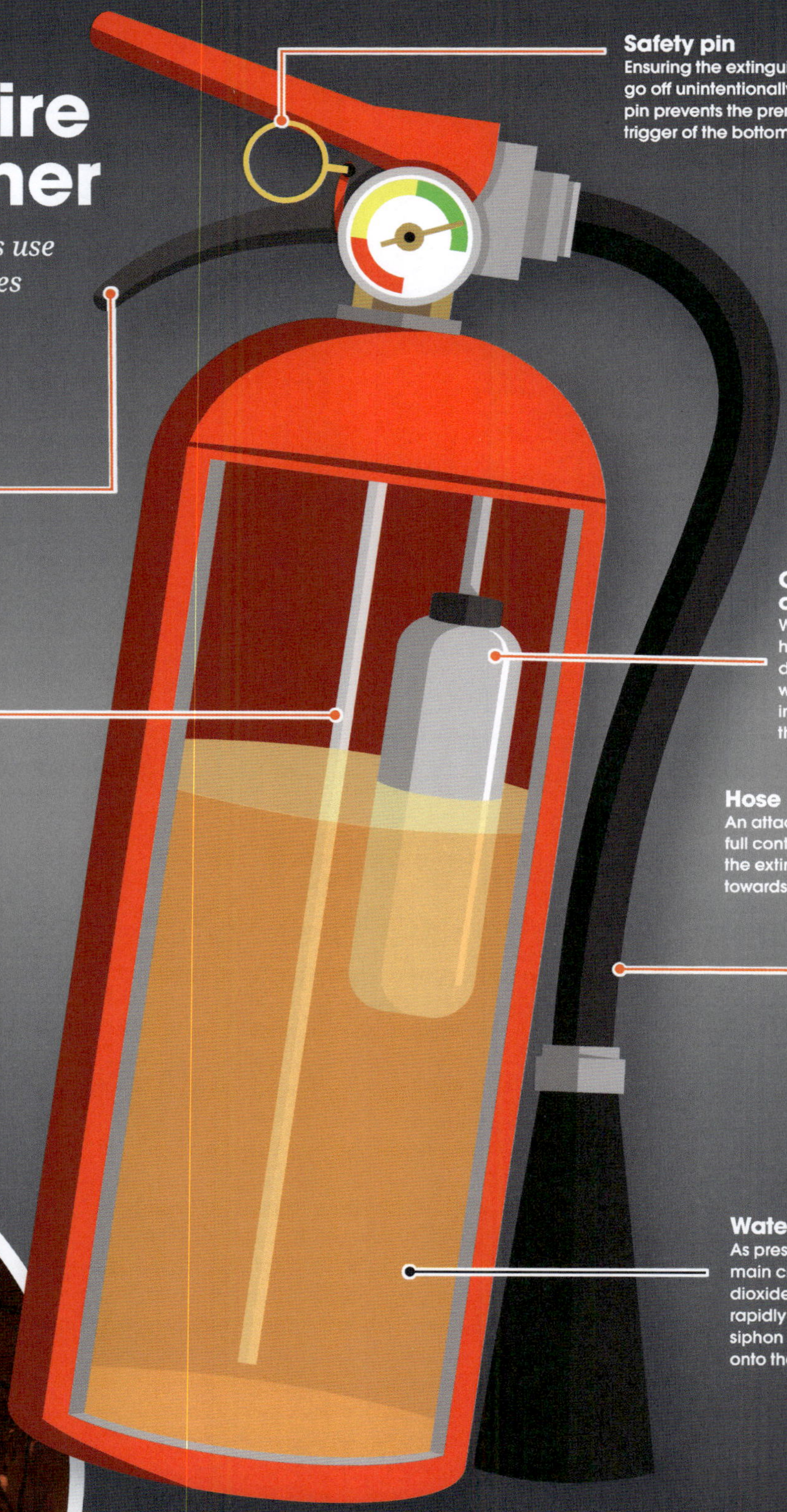

Safety pin
Ensuring the extinguisher doesn't go off unintentionally, a safety pin prevents the premature trigger of the bottom handle.

Handle
To activate the extinguisher and release the water within, the handles needs to be squeezed together.

Carbon dioxide canister
When triggered by the handles, the carbon dioxide canister is opened, which releases pressure into the main chamber of the extinguisher.

Siphon tube
When the carbon dioxide canister is activated, the extinguisher's water content is rapidly driven up the siphon tube for release.

Hose
An attached hose allows full control over directing the extinguisher's contents towards the fire.

Water
As pressure builds within the main chamber by the carbon dioxide canister, water is rapidly expelled up the siphon tube, out the hose and onto the fire to smother it.

When heated to around 150 degrees Celsius, wood begins to decompose, releasing volatile chemicals as smoke

or more than five months, more than 11 million ectares of Australian bush were destroyed by vildfires. The ignition source could have been lightning strike or even arson. But whatever was, these fires raged so fiercely because of he dry environmental conditions in Australia. ointing the finger towards climate change and he extreme heat the country had experienced hroughout the previous year, Australia's forests nd bush had become the perfect kindling for he fire to exploit.

As a forest fire rages, hot air rises upwards nd cooler oxygen-rich air is dragged to the ottom, which is the same as when a candle ame burns. The air's motion is barely oticeable in the tiny flame of a candle, but vhen a fire is burning across hectares, this aseous movement can generate winds of nore than 160 kilometres per hour. With he assistance of strong wind speeds, embers roduced by fires leap from one wildfire and ll to another area of forest, igniting a new laze. Without widespread rainfall to extinguish he travelling flames, Australia experienced the vorst fire season on record.

The hottest point of a gas flame – the tip – can reach 1900 Celsius. It's a blue colour because the increased oxygen supply creates a more efficient and energetic combustion that ionizes the gas molecules

Fire in space

Lighting a match inside a giant, oxygen-filled research centre floating in space doesn't sound like the safest idea. However, in 2012, astronauts aboard the International Space Station (ISS) did just that. With a lack of oxygen throughout space, the only way to see how fire reacted in microgravity was to strike a match aboard the ISS. The teardrop shape of a flame on Earth is due to the uptake of oxygen as it burns. However, in microgravity, the same rules don't apply. Without an abundance of rising oxygen, space flames are shaped merely by the static air around them. This creates a spherical flame as the chemical reactions creating light occur all around the match.

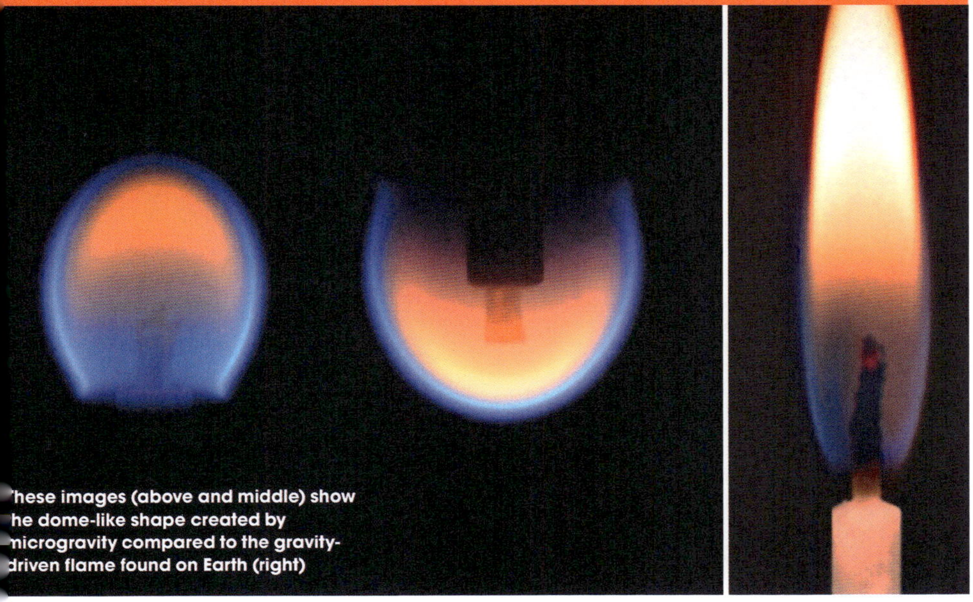

These images (above and middle) show the dome-like shape created by microgravity compared to the gravity-driven flame found on Earth (right)

Prescribed burning can be used to create a finish line for raging wildfires

Fighting fire with fire

When battling blazing wildfires it might seem strange to introduce more fire in an attempt to quash the already-raging flames. However, prescribed burns may help to relieve the painful effects of wildfires. In a strategic endeavour to limit the advancement of wildfires, a human-made wall of fire can be drawn around them in the hopes of starving them of fuel. By purposely burning strips of woodland before wildfires can reach them, a prescribed fire could potentially stop the spread. This isn't a guarantee, and although controlled by firefighters, there is a risk the prescribed fire could become uncontrollable, or that floating embers could ignite new wildfires.

How fireworks work

Discover the science behind the glitter and sparkle of these pyrotechnics

Whether it's the Fourth of July, Guy Fawkes Night or New Year's Eve, you're sure to see the sparking light of nearby fireworks, followed quickly by their boisterous booms. But igniting colourful chemicals and shooting them into the sky isn't a new tradition by any means. The first fireworks were hurtling into the sky in around 600 BCE in ancient China. Ancient alchemists discovered that mixing potassium nitrate, sulphur and charcoal produced a black powder that's now referred to as the 'first gunpowder'. When this powder was stuffed into a bamboo stick and ignited, it created the first firework propulsion system. Early fireworks were used to celebrate weddings, births and even to ward off evil spirits.

Modern fireworks still use black powder, but now incorporate small pellets of packed minerals, called stars, to produce colourful explosions. When the chemicals in the stars are heated, the electrons making up their atomic structures are excited and release energy in the form of different wavelengths of light. For example, when a copper-filled star ignites it burns a brilliant blue colour. Stars are held within a shell, often concealed within a rocket, and are placed in a launching tube called a mortar. A pile of black powder beneath the shell is ignited, and its combustion generates enough force to send the shell flying into the sky before another charge inside the shell combusts, igniting the shells, and the stars within them. If the stars inside are arranged in a particular orientation or design, the erupting fireworks will also resemble that design. So if the stars are arranged to look like a heart in the shell, the resulting firework will also resemble a heart.

Shells come in different shapes and sizes. In the US shells are typically spherical, whereas in Europe shells are more cylindrical and stack layers of different stars on top of each other, separated by a portion of black powder, to ignite them in stages. Of course, the fireworks aren't purely visual – there's also a range of hissing, crackling and booming sounds that come with a good firework display. These auditory elements of a firework aren't necessarily natural byproducts of a firework explosion... some are purposefully put into the firework shell. Minerals such as bismuth create crackling sounds, while potassium salts make fireworks whistle as they travel.

AI-designed displays

Like in many other industries, artificial intelligence (AI) has found its way into the toolset of firework display designers. Along with being able to simulate a digital display, AI is being used to gather data from displays around the world to choreograph new and unique experiences. AI is capable of precisely calculating the optimal timing, position and safety of fireworks to produce a display that's not only visually exciting, but one that is perfectly synchronised with any accompanying music. It can even be used to respond to changes in the weather in real time, such as adjusting the position of a launching mortar to account for changing wind speed.

AI is often used to calculate the perfect combinations of fireworks for ideal displays

Creating colours

The elements that give fireworks their colours

Colour	Element(s)
GREEN	BARIUM
RED	STRONTIUM
YELLOW	SODIUM
BLUE	COPPER
SILVER	TITANIUM, ALUMINIUM AND MAGNESIUM

Inside the shell

How fireworks are sent flying and result in predictable explosions

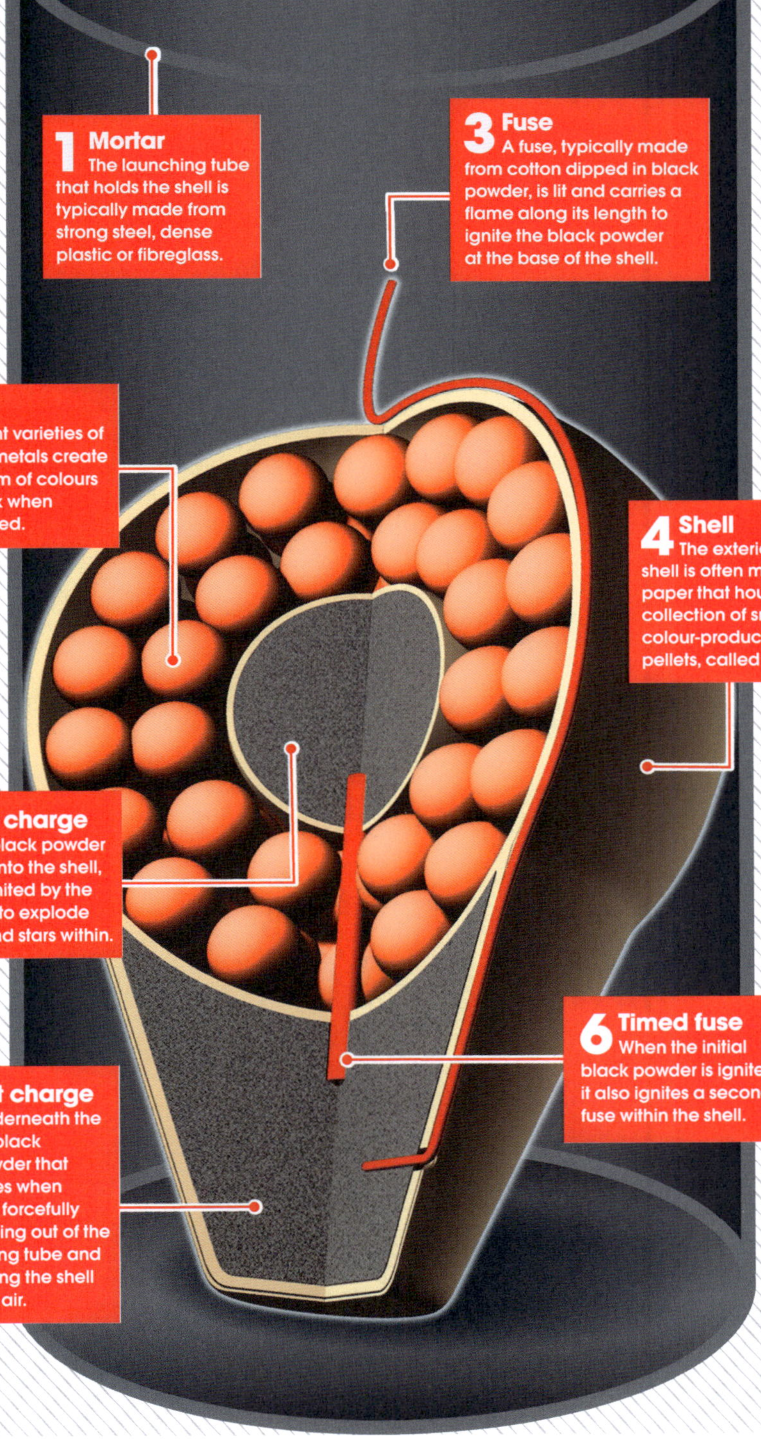

A firework mortar with fuses, ready to be filled with black powder and shells

5 Fun firework patterns

1 Willow
As well as coloured salt, a large amount of charcoal is put into the stars of a willow firework. When the stars explode, the charcoal burns in trails as it falls.

2 Palm
When larger grains of minerals are used within the shell's stars, it takes longer for them to combust, creating a palm tree-like effect.

3 Strobe
The composition of the compounds used in these dazzling fireworks produces a strobing effect, caused by chemical byproducts of the combustion that re-ignite following the initial explosion.

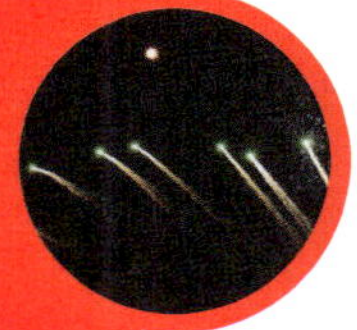

4 Comet
A large amount of charcoal, along with titanium or aluminium, is used to produce long gold and silver tails radiating from a comet firework.

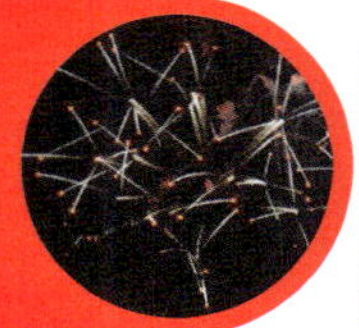

5 Crossette
Unlike a typical firework star, the stars of a crossette firework have a four-sided cavity filled with a burst charge and capped at one end. When the firework explodes, the star separates in four different directions.

Liquid nitrogen

The substance that's deadly enough to stop a Terminator

5 Uses of liquid nitrogen

1 Making ice cream
Ice cream frozen with liquid nitrogen is smooth, creamy, light in texture and a litre can be made in 30 seconds. Fans of this cool technique include world-class chef, Heston Blumenthal.

2 Preserving specimens
Rapid freezing can be used to preserve tissue such as blood and bone marrow. Other items that can also be stored in liquid nitrogen include animal embryos, bacteria, and fungi.

3 Removing warts
Freezing off warts and verrucas using liquid nitrogen is a common, quick procedure that's simple enough to do in a doctor's office. The extreme cold destroys the unwanted skin cells.

4 Recycling tyres
Rubber tyres are built to be durable, making them hard to destroy. Freezing with liquid nitrogen makes them brittle, so they can be smashed into 'crumbs' for uses like building running tracks.

5 Powering cars
Liquid nitrogen could be a futuristic fuel. As it rapidly changes from a liquid to a gas, the resulting gas flow could power a turbine and turn car wheels.

Liquid nitrogen can be used to make silky ice cream

Nitrogen gas makes up around 78 per cent of the Earth's atmosphere – however liquid nitrogen is uncommon. This is because nitrogen gas has to be cooled to -195.8°C before it is transformed into a completely colourless liquid. By contrast, water becomes a liquid below 100°C.

Liquid nitrogen absorbs lots of heat when it evaporates so it's useful for rapidly cooling things like food. Quick cooling means large ice crystals don't have time to grow, burst cells and damage food quality. Since nitrogen is inert – it reacts with few chemicals – it's safe to immerse food and it displaces the oxygen that harmful bacteria need to grow.

When liquid nitrogen evaporates, it creates around 700 times its volume in gas and pushes aside the air surrounding it. If there's not enough ventilation in a room, the nitrogen can displace all the oxygen and suffocate anyone inside.

A great way to stop your enemies in their tracks, just like in *Terminator 2*

"When light passes through the crystals, it refracts, allowing it to pass through the filters at different angles"

Crystallised alcohol

The hidden beauty of our favourite alcoholic drinks

If you leave a drop of an alcoholic beverage to dry out, the water and alcohol will eventually evaporate to leave behind crystallised sugar. If you then look at this sugar through a polarising microscope, you will e a pattern of bright colours as light refracts through the crystals. The effect is created using two polarising filters, one between the crystals and the light underneath them, and another positioned at a 90-degree angle from it, between the crystals and the microscope lens above them. As these filters force light waves to oscillate in one direction, rather than all different directions as they would normally, the two polarising filters should block the light completely. But when the light passes through the crystals it refracts, allowing it to pass through the filters at different angles so we see lots of vibrant colours. Geologists use the same technique to study the structure and composition of rocks.

The science of distillation

The revolutionary ancient discovery that enables us to make gasoline from crude oil and vodka from potatoes

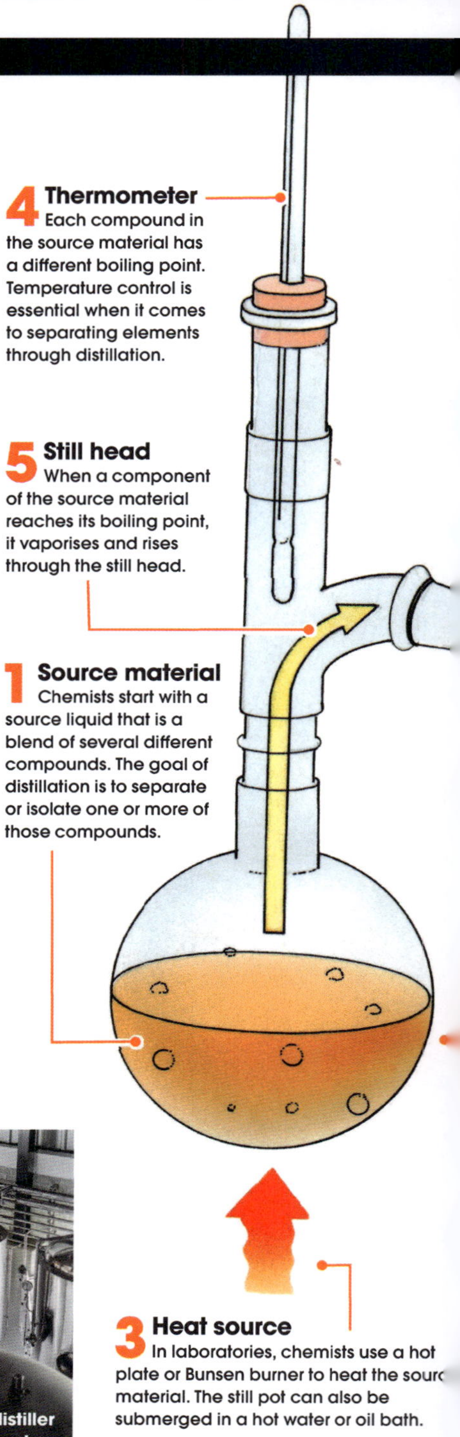

In many ways, during the Middle Ages, distillation was the alchemist's greatest tool. Through controlled heating of a liquid or solid source material, the alchemist could draw out and isolate pure vapours that were cooled, condensed into liquid and employed in the mystical medicinal research. Distillation also helped the ancient Greeks to purify fresh drinking water from the sea. It drew pungent oils out of flowers and herbs for perfumes and early medicines. And it took humble beer and homemade wine and refined them into finely crafted, high-octane spirits that earned the name aqua vitae, or the water of life.

Today, distillation remains a critical chemical process for isolating pure compounds from a mixed solution. The science of distillation is rooted in the fact that every molecular compound has its own unique boiling point and dew point. The boiling point is reached when the vapour pressure of a liquid equals the atmospheric pressure above it, allowing molecules within the liquid to freely escape as vapour. The dew point is the temperature at which the vapour cools enough to condense and return to a liquid state.

Compounds have unique boiling points, so can be vaporised out of a mixed solution at different temperatures. First, you slowly heat the solution to the boiling point of the compound you want to isolate. The vapour that rises from the mixture will still contain a blend of the compounds in the solution, but a far greater concentration of the desired compound. The rising vapour is channelled into a condensing tube, where it is cooled by a circulating water bath and then condensed back into a purified liquid. Distillation is taught in school labs, but industrial-strength distillation supplies the world with everything from whisky and perfume through to gasoline and liquid nitrogen, plus a lot more besides.

A cutting-edge distiller in a modern laboratory

Making alcohol

The alcohol found in wine, beer and hard liquor is really ethanol, the same stuff that powers alternative-fuel vehicles. To make ethanol (alcohol), all you need is a sugary (high carbohydrate) feedstock like grapes, corn or malted barley, plus water and yeast. Yeasts are microscopic organisms that metabolise the carbohydrates in the feedstock and produce two by-products: ethanol and carbon dioxide.

The carbon dioxide produced by yeast is what makes bread dough rise. To encourage ethanol production, brewers cover vats of fermenting material and allow only the carbon dioxide to escape. Wine, beer and other fermented brews – like the grain mash that's used to make whisky – can be distilled to concentrate their alcohol content, creating more potent spirits like gin, vodka, brandy and rum.

Fractional distillation

In simple distillation, a single distillate is vaporised out of a solution and condensed in a single batch. Fractional distillation, meanwhile, is a technique for carefully separating multiple different components of a solution in a continuous process. The separation process occurs in a fractionating column that sits above the heated source material. The column is packed with tiny glass beads and contains multiple levels of condensing trays. An important characteristic of the fractionating column is that internal temperatures are hotter at the bottom and cooler at the top. As each component of the source material reaches its boiling point, it vaporises upward through the column. The components with the lowest boiling point rise higher in the column before cooling enough to condense back into liquid. The glass beads provide extra surface area to promote condensation, and the condensing trays collect each liquid component, which can then be siphoned out. Crude oil is separated into its useful by-products – eg kerosene, diesel fuel, natural gas and more – in towering fractionating columns.

Simple distillation

How a single compound can be extracted from a mixed solution

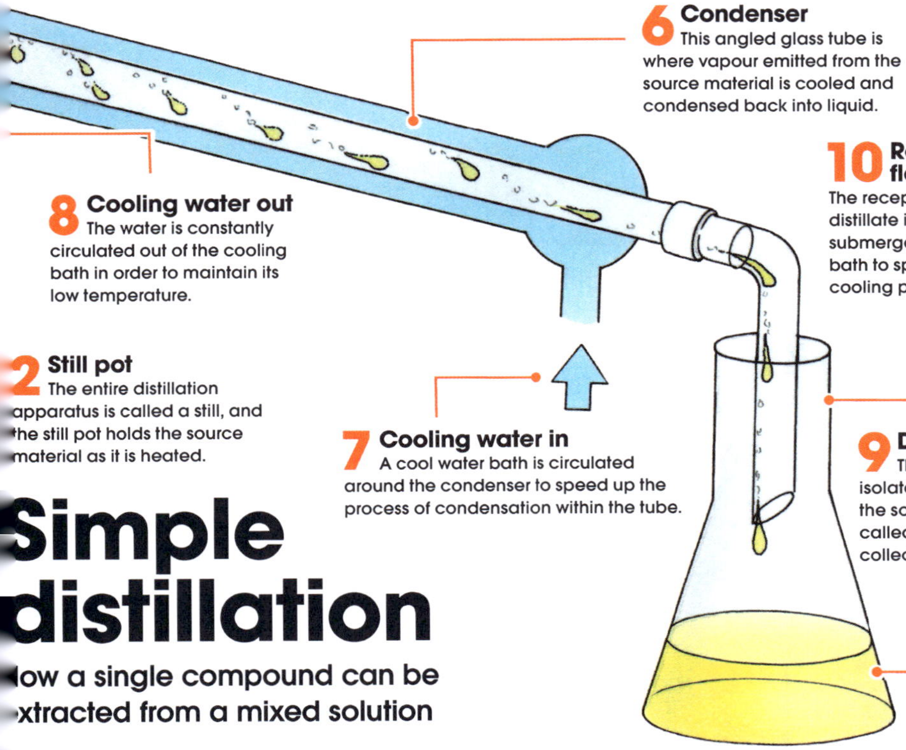

6 Condenser This angled glass tube is where vapour emitted from the source material is cooled and condensed back into liquid.

8 Cooling water out The water is constantly circulated out of the cooling bath in order to maintain its low temperature.

10 Receiving flask The receptacle for the distillate is sometimes submerged in a water bath to speed up the cooling process.

2 Still pot The entire distillation apparatus is called a still, and the still pot holds the source material as it is heated.

7 Cooling water in A cool water bath is circulated around the condenser to speed up the process of condensation within the tube.

9 Distillate This is where the isolated component of the source material, called the distillate, is collected and cooled.

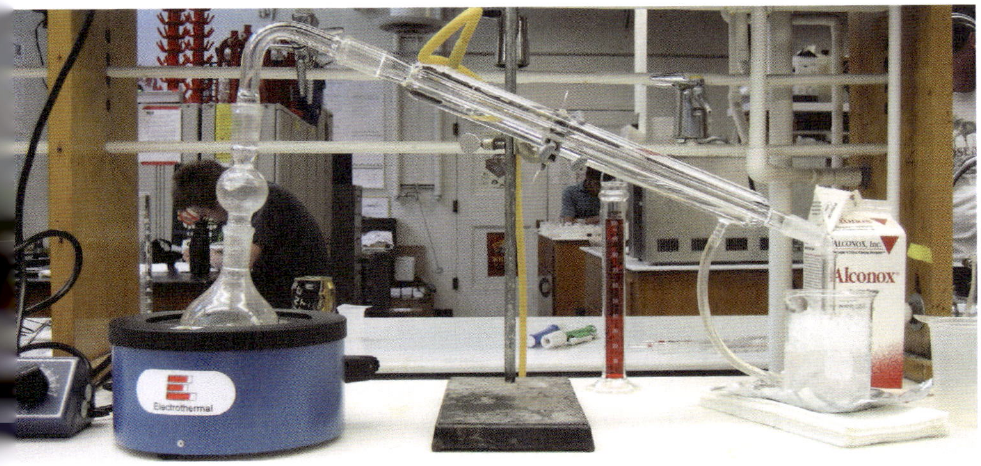

5 Uses of distillation

1 Distilled water

By boiling water and condensing its vapour, you achieve a purified water with lower levels of trace minerals and other impurities. Originally, distillation was used to make fresh drinking water from seawater, but now the purified liquid is used in car batteries and other chemical-electrical systems where the minerals found in tap water would disrupt the reaction.

2 Alcoholic spirits

Beer and wine pack a pleasant alcoholic punch, but the alcohol content of such beverages is never more than 20 per cent by volume. Taking their cue from Arab alchemists, European brewers from the 14th century distilled beer and wine to concentrate ethanol content, creating potent spirits like brandy and gin.

3 Oil refining

The crude oil that we pump out of the ground is absolutely useless as a fuel. First, it must be refined through fractional distillation and separated into its component parts: diesel fuel, kerosene, gasoline and natural gas. Crude oil is heated to a ferocious 600˚C and its vapours ascend through a cooling column of air. As each component of crude oil reaches its boiling point, it condenses and flows out of the distillation column.

4 Separation of air

Air can be separated into its component elements – mainly nitrogen, oxygen and argon – via a process called cryogenic distillation. In cryogenic distillation, gaseous air is first pressurised and then cooled to temperatures approaching -200˚C, at which point all of its components condense and separate into purified liquid nitrogen, oxygen, argon and so on.

5 Perfume making

Persian alchemist Ibn Sina (aka Avicenna) is credited with first distilling essential oils to create rose water. Today, compounds come from flowers, herbs, resins and even some animal-based substances like ambergris, a musky grey lump regurgitated by sperm whales.

The science behind food

Take a look at the chemistry that goes on in the kitchen when we cook our food

Some of the most interesting kitchen chemistry can be observed when baking. Taking four basic ingredients – flour, fat, sugar and eggs – and subtly altering their cooking chemistry can transform them into airy cakes, chewy cookies or flaky pastries.

Leavening, or raising, agents introduce bubbles of air. As the air bubbles are heated, the gas that they contain expands, causing cakes, breads and soufflés to rise. These air bubbles can be made in one of two ways. Chemical raising agents, like baking powder and bicarbonate of soda, react with water to form carbon dioxide gas. This reaction occurs very rapidly and the quantity of raising agent must be carefully adjusted – too much and the bubbles will become large and burst, too little and the density of the cake mixture will prevent any bubble formation at all.

For a slower rise with added flavour, baker's yeast (Saccharomyces cerevisiae) is often used. Yeast is a single-celled organism of the fungi family. At first, the yeast respires aerobically – using oxygen – creating bubbles of carbon dioxide. When the oxygen runs out, the yeast begins to make ethanol by fermentation, much like in brewing beer, but any alcohol formed in the bread dough evaporates in the oven.

Making bubbles is one thing, but getting them to remain intact requires more clever chemistry. Bread is most often made from wheat flour, which contains starch granules surrounded by two important proteins: glutenin and gliadin. When mixed with water and kneaded, the glutenin cross-links to form networks with gliadin, making a new stretchy protein: gluten. Gluten is a 'super-protein', or protein complex, which behaves like elastic, forming stretchy bridges that hold the starch molecules together. The key to light, fluffy bread lies in creating lots of tiny elastic bubbles; the more the dough is kneaded and stretched, the stronger the gluten network. Eggs act in a similar way to the gluten in flour, providing a protein-binding agent that supports air bubbles and holds cakes together.

Unlike bread, pastries need to be 'short' and crumbly, so bakers try to minimise gluten

Why cooked meat tastes better than raw

The brown colour of seared meat, toasted bread and roasted coffee beans is down to a chemical process called the Maillard reaction. When foods are heated, amino acids (the building blocks of protein) react with sugars to create hundreds of flavour compounds. Depending on the types and quantities of the various amino acids in the food, different combinations of flavours will be produced, as this diagram shows...

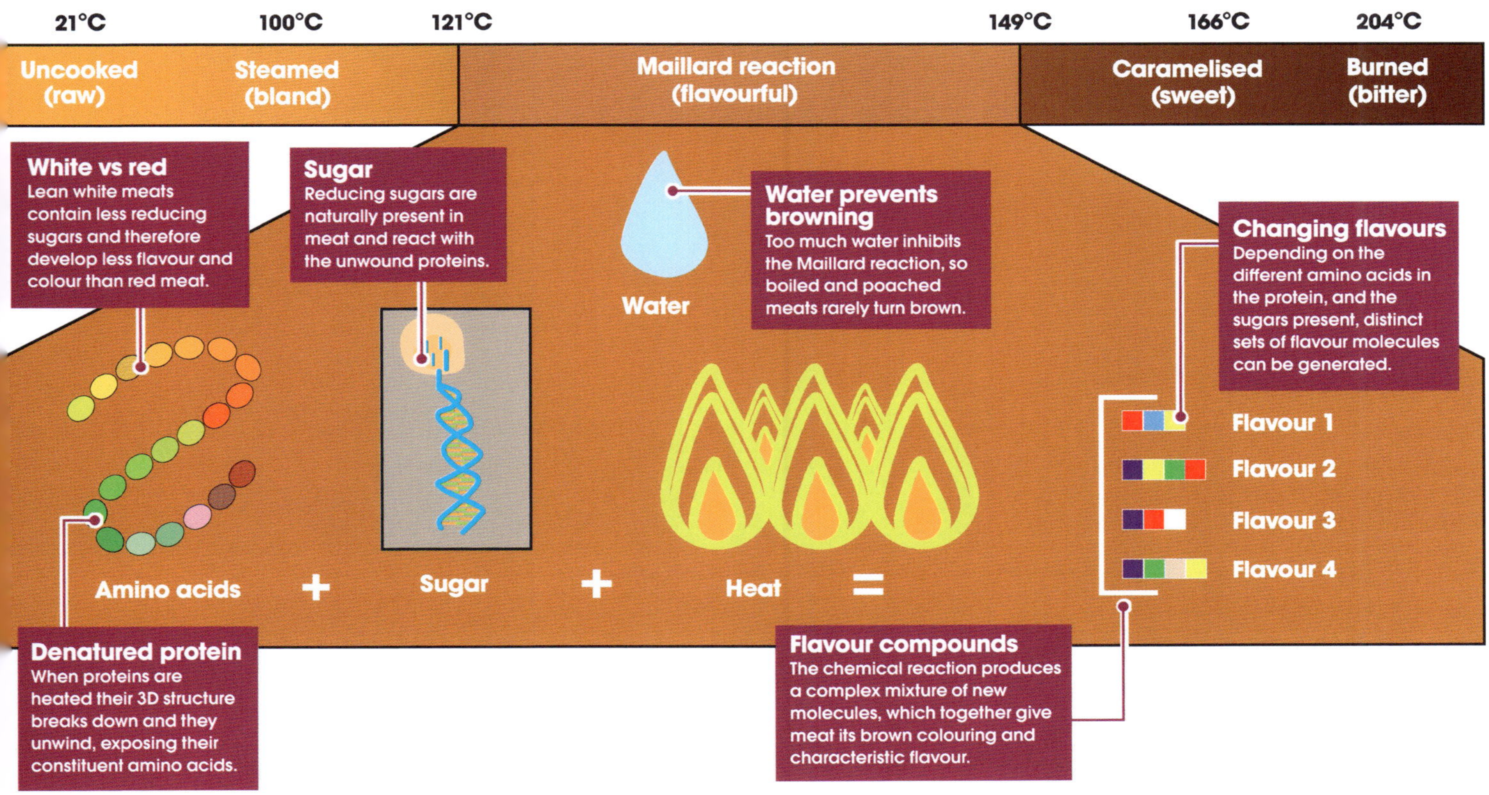

roduction, which would lead to a rubbery xture. This is done by first rubbing butter into e flour, coating the starch molecules with a yer of fat, which helps prevent glutenin and iadin from coming into contact with water.

The texture of baked goods can also be altered sing sugar. When sugar is beaten with butter, e sharp edges of the sugar crystals allow ny air bubbles to form – turning the mixture pale, creamy yellow colour. These bubbles xpand in the same way as the ones created by ising agents, contributing to the light texture cakes. For the denser consistency of cookies, elted fats and oils are often used because the ndency for bubbles to form next to the sugar ystals is reduced.

Sugar also draws in moisture from the air, hich can have a significant effect on the water ntent of baked goods. Brown sugar attracts ore water than white, and finely ground sugars attract more water than the granulated variety. Experimenting with the type of sugar used in a recipe will alter the final moisture content, and therefore the texture.

> **"Experimenting with the type of sugar used in a recipe will alter the final moisture content, and the texture"**

Chemistry isn't just limited to baking though. Chemical reactions define the taste of meat – which is around 70 per cent water, with the remainder being mostly protein and fat. Depending on the cut, meat contains a variable amount of collagen – a fibrous protein in the skin, tendons and connective tissue. The higher the collagen content, the tougher the meat is.

More expensive cuts and meat from younger animals contain little collagen and are able to be cooked rapidly. The muscle protein myosin denatures (breaks downs) at low temperatures – ie 50 degrees Celsius – which causes it to begin forming cross-links. These lend some support to the structure of meat. At this stage, water molecules between the proteins start to leak out, but the meat will still remain juicy and tender. Once the temperature reaches 60 degrees Celsius, the red pigment in muscle – myoglobin – denatures to form a hemichrome. It is this that gives cooked red meat its brown-grey colour.

Further heating causes the collagen to shrink and contract, forcing water out and turning the meat from juicy and tender to chewy and dry. If the temperature is raised further – to, say, 70 degrees Celsius – the meat continues to toughen, but the collagen itself dissolves to form gelatine. Although the fibres of meat are more brittle, the gelatine acts as a lubricant, giving slow-cooked meat its soft texture.

Heat isn't the only way to break down collagen, and meat can be physically or chemically tenderised. Marinades work by using common culinary chemicals to interfere with the bonds between collagen strands – these can range from acids such as lemon juice to enzymes like bromelain.

Another great example of kitchen chemistry is the emulsion process. Oil and water don't mix, but to make sauces like mayonnaise and béchamel, a cook needs a way to bring them together. When oil and water are combined, the oil floats on top, forming an interface with the water that has high surface tension. In order to break this tension, mechanical shearing can be used – by shaking the container the oil breaks down into smaller and smaller bubbles, which disperse into the water. However, this is only a temporary emulsion, and after a while the oil and water will separate.

Mayonnaise contains watery egg yolks and fatty butter, which must mix to form a smooth, white paste in a permanent emulsion. Egg yolks contain an emulsifier called lecithin, which dissolves in both fat and water, essentially forming bridges between the yolk and the butter and holding the mayonnaise emulsion in a stable structure. Flour can be used in a similar way in white sauces like béchamel; the fine powder helps to bind the butter to the liquid.

The flavour of food is determined by its combination of volatile components that get into the air and interact with sensory neurons in the nose. Each food may have hundreds of these molecules, but scientists studying flavour combinations have seen that if one of them matches, foods are likely to go together. The technique is being used to predict unlikely food partners.

Molecular gastronomy takes the science of cooking to the next level. Looking at food from a purely physical and chemical perspective, a host of chefs and scientists are coming together to identify new flavour combinations and cooking techniques based on science. Using liquid nitrogen, syringes, centrifuges and ultrasound machines, we are starting to reinvent the way we cook. As the founding father of this scientific discipline, Nicholas Kurti, said: "It's a sad reflection on civilisation that while we can and do measure the temperature in the atmosphere of Venus we do not know what goes on inside our soufflés".

The science behind soufflés

Approach soufflé making like a lab experiment and you'll get great results every time

1 Fat is your enemy Fat pops bubbles, so you must minimise any contact with the egg whites. →

2 Use fresh eggs Old eggs may whip up faster, but the bubbles are larger and less stable. →

3 It's no yolk Ensure that there are no traces of the fatty yolk contaminating the white. ↓

4 The right bowl Plastics contain fat-like molecules, so use a glass or metal bowl for stirring up the mixture. ↓

5 Perfect peaks Beat the egg whites until they form beautiful stiff, foamy peaks. ←

6 Firm filling This supports the weight of the bubbles so make it thick. ←

7 Greased dishes If the soufflé gets stuck, the bubbles will pop. ←

8 Be gentle Folding the filling in should take less than a minute – be very gentle as you do it. ↑

9 Do not disturb Turn the oven's fan off and don't open the door until it's nearly done.

Strange flavour combinations

Some seemingly odd food pairings are surprisingly good, others are just plain disgusting – but why?

Chocolate and salt
Salt actually helps the cells on your tongue to sense the presence of sugar, so it makes chocolate taste even sweeter.

 +  = YUM

Peanut butter and apple
Peanut butter makes up for what an apple lacks in salt and fat, while the apple cuts through the spread's richness and stickiness.

 +  = YUM

Citrus fruit and milk
The acid found in citrus fruit causes milk to separate and curdle – essentially the first step in making cheese. Not appetising.

 +  = YUCK

Chilli powder and fruit
The compound capsaicin present in chilli has two effects: it enhances our sense of smell and also heightens our perception of sweetness.

 +  = YUM

Coffee and olives
We evolved to associate bitterness with poison, so combining too many bitter flavours is often an unpleasant experience.

 +  = YUCK

A matter of taste

How do we differentiate the flavours of food?

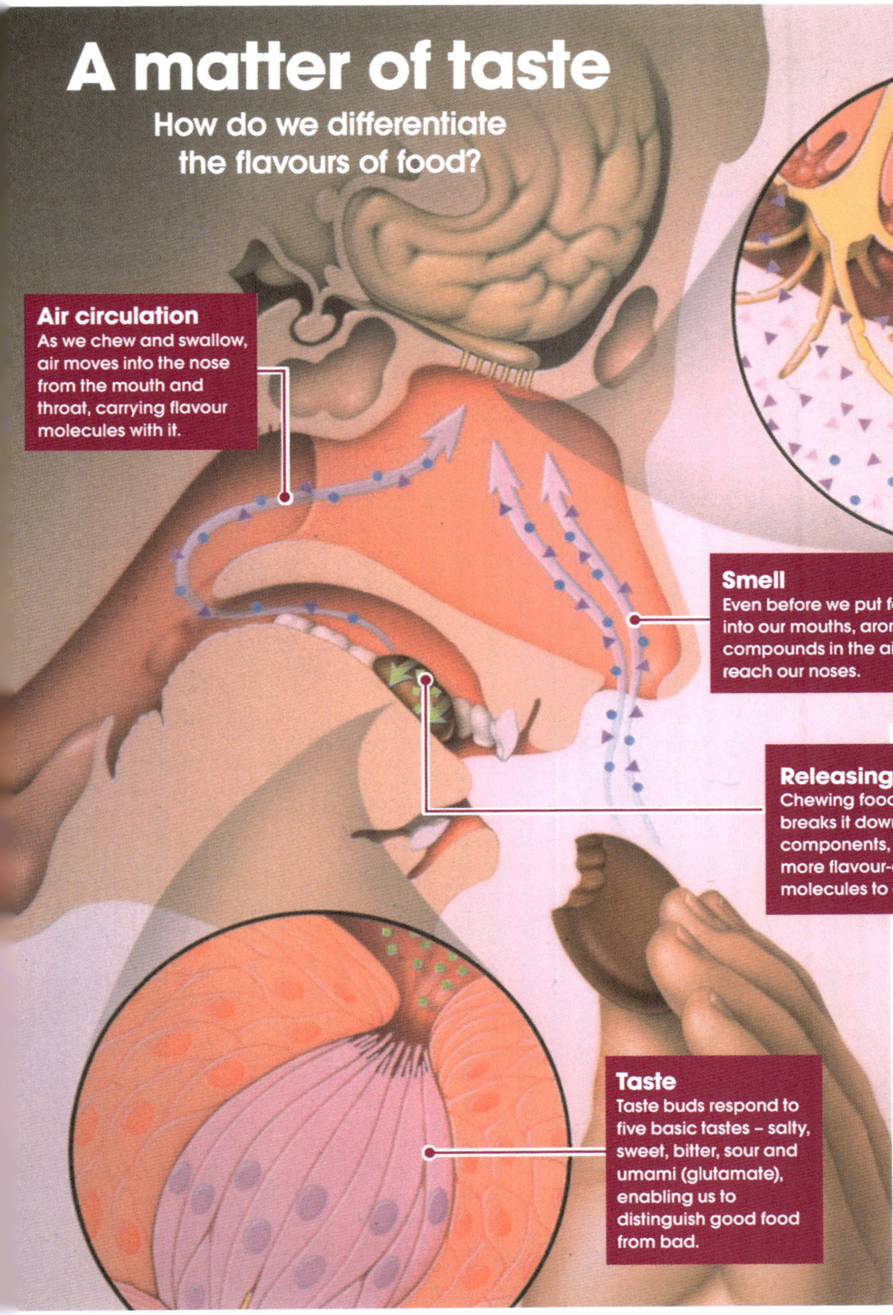

;an food make us happy?

ɘople often report cravings for particular ods, and that eating certain meals makes em happy. As a species, we evolved to ake eating a pleasant experience, ıcouraging us to seek out high-calorie food sustain ourselves when food was scarce. e human brain has developed reward ıthways associated with eating fat and gar, which release mood-enhancing neurotransmitters, like dopamine and endorphins. Probably the most-studied example is chocolate, which contains phenylethylamine and this affects the body's opioid production. Comfort food, on the other hand, works more psychologically, and the pleasant feelings that it induces are often linked to sight, smell and taste, which can trigger a sense of nostalgia.

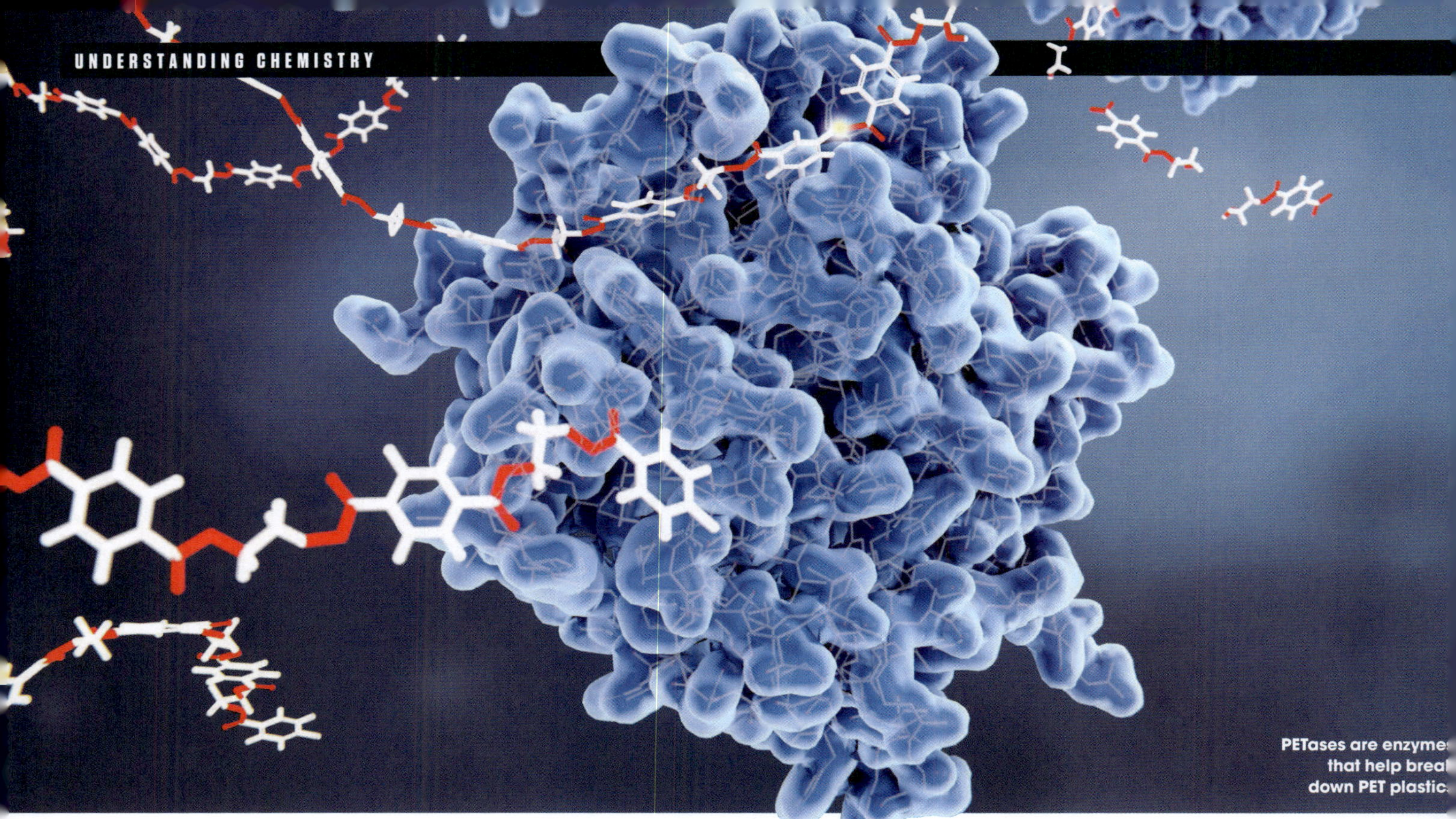

PETases are enzymes that help break down PET plastic

Enzymes

Meet the molecules that regulate the biochemical reactions in our bodies

Enzymes are often called 'biological catalysts', and their job is to speed up chemical reactions. You are full of dissolved chemicals with the potential to come together or separate to form the biological building blocks that you need to stay alive, but the reactions happen too slowly on their own.

Enzymes are molecules with 'active sites' that lock on to other molecules, bringing them close together so that they can react, or bending their structures so that they can combine or break apart more easily. The enzymes themselves do not actually get involved in the reactions; they just help them to happen faster.

Some of the most well-known enzymes are the ones in your digestive system. These are important for breaking down the molecules in your food. However, these aren't the only enzymes in your body. There are others responsible for building molecules, snipping molecules, tidying up when molecules are no longer needed, and some even help to destroy invading pathogens.

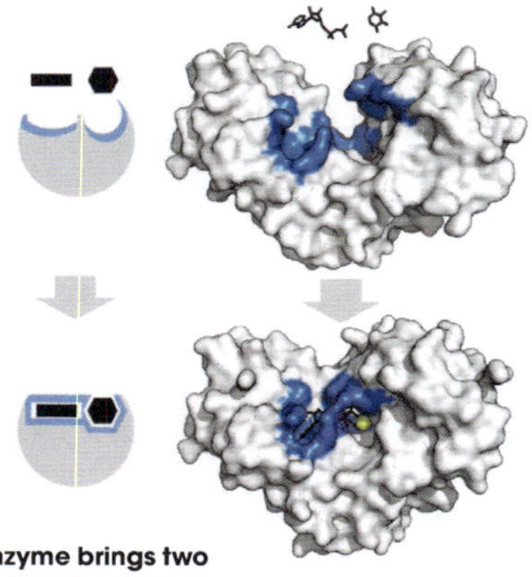

This enzyme brings two molecules close together so that they can react

Background

Your body is like a walking, talking biochemical reactor, and enzymes are the catalysts. Enzymes are molecules that boost the rate of chemical reactions by bringing other molecules together so that they can react.

Industrial enzymes

Enzymes are used for many different processes in agriculture, cosmetics, the food industry and more.

Perhaps their most well-known application is in biological washing detergents. These contain digestive enzymes (either one or a mixture of proteases, amylases and lipases) to help break down stains when washing clothes at lower temperatures. Enzymes can also be used when brewing beer; amylase will help turn starch into maltose and glucose, which yeast then feed on to start the alcoholic fermentation process.

Lipases are naturally found in unpasteurised milk, and can be used in the dairy industry to help produce cheeses. By breaking down fat globules into fatty acids, lipases can help improve the flavour of a cheese and play a role in the ripening process. Enzymes can only be used in certain conditions, though. Extremes of temperature, pH and other chemicals can cause them to 'denature' – changing their shapes so the active sites no longer work.

Digestive enzymes

Find out which proteins help us break down our food

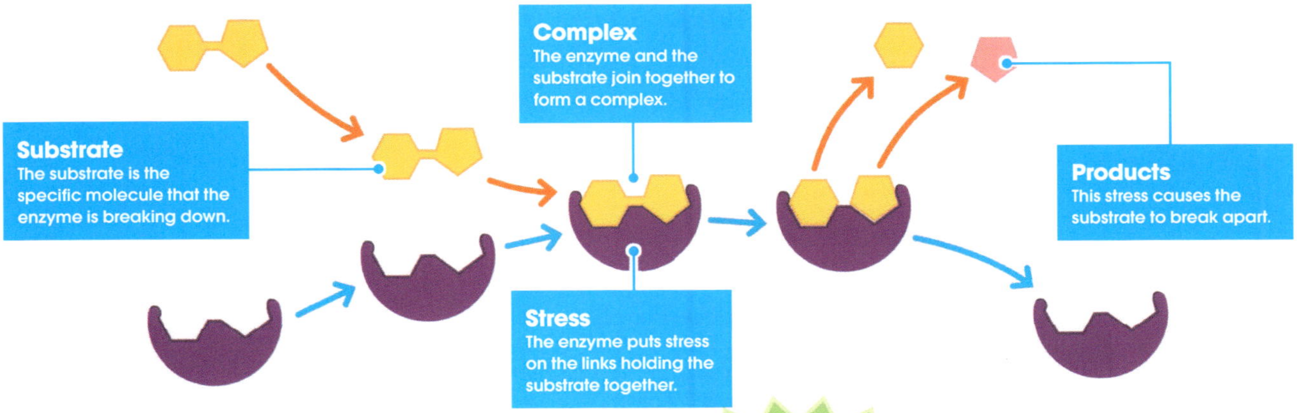

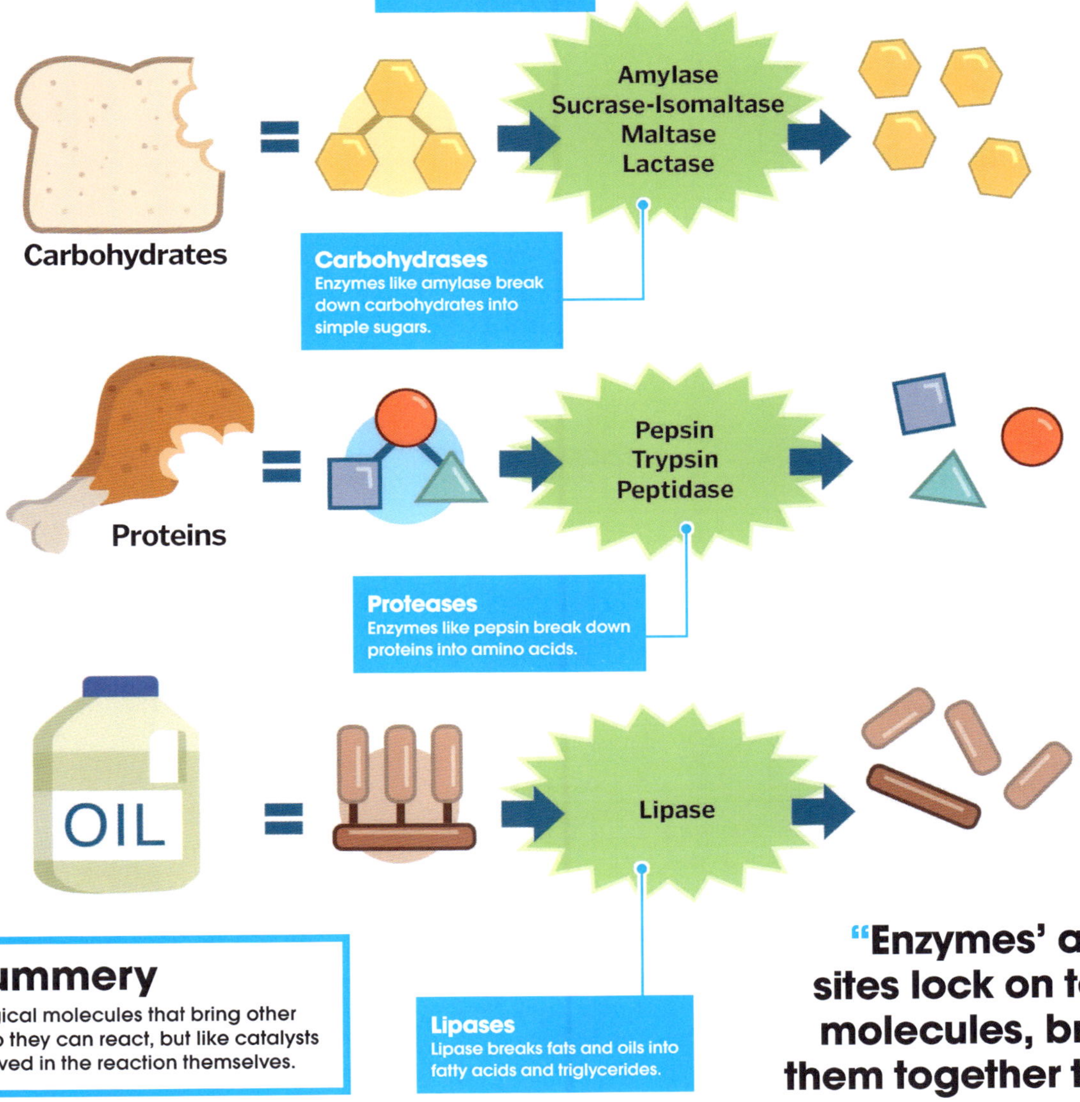

Summery

Enzymes are biological molecules that bring other molecules together so they can react, but like catalysts they don't get involved in the reaction themselves.

"Enzymes' active sites lock on to other molecules, bringing them together to react"

Creative culinary science

Discover the chemistry behind the delicious dishes created by the top chefs trained in molecular gastronomy

Gels

When experimenting with textures, chefs are able to turn liquids into solids by adding a gelling agent called a hydrocolloid. Common hydrocolloids used in cooking include gelatine, agar-agar and sodium alginate and are made of tightly-wound coils of long protein or carbohydrate molecules. When added to liquid, the molecules evenly disperse and then heat is applied to cause them to hydrate and unfurl. As the mixture cools back down again, the long strands form a crosslinked network that traps the liquid molecules and forms a solid shape. The result is a firm gel or jelly packed full of the flavour of the original liquid.

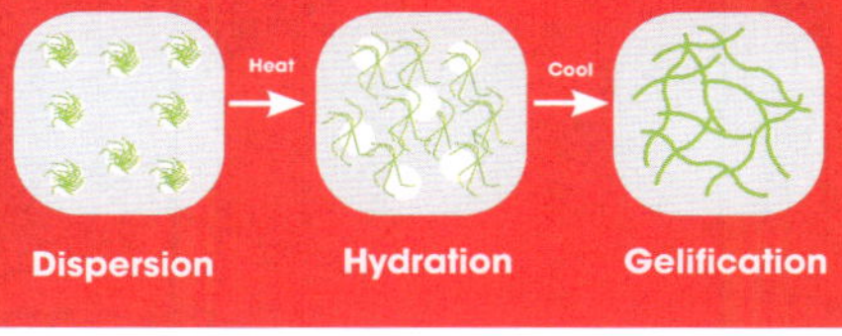

Liquid nitrogen

To introduce some drama to their food presentation, chefs can garnish their drinks and dishes with a dash of liquid nitrogen to produce a magical smoke effect.

Nitrogen is normally a harmless, odourless and tasteless gas and is present in the air we breathe, but when cooled to below -196 degrees Celsius it becomes a liquid. If it is exposed to warmer temperatures again, it quickly boils and evaporates, condensing the moisture in the surrounding air to create a dense fog.

Liquid nitrogen isn't just useful for decoration, though, as it can also be used to make delicious frozen desserts. When making ice cream, the supercooled liquid will freeze the ingredients instantly. This prevents ice crystals from forming so that the finished product is incredibly smooth and creamy.

Nitrogen must be handled with care as a liquid but is harmless as a gas

Spherification

To add an explosion of flavour to a dish or cocktail, chefs can use clever chemistry to create spheres of flavoured liquid. The liquid is contained inside a thin gel membrane that when put under slight pressure in your mouth, bursts open to reveal the delicious contents within. The spheres can be created using one of two spherification techniques. Basic spherification has the benefit of creating an extremely thin gel membrane that is almost imperceptible in your mouth, but this will not work if the liquid has a high acidity. The gel also keeps forming after the liquid is removed from the bath, so the sphere must be served immediately, before it becomes a solid gel ball. Reverse spherification is best used for liquids with a high calcium or alcohol content, and although it creates a thicker membrane, the finished sphere is longer lasting as the gel stops forming as soon as it is removed from the bath.

Spherification can be used to create small, caviar-like balls or large domes

Basic spherification

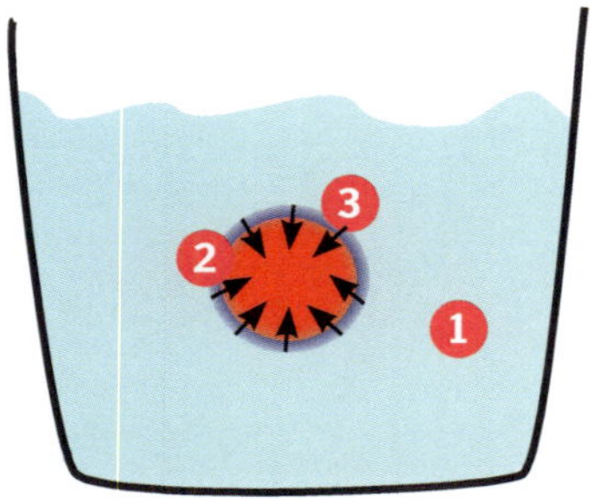

1 Sodium alginate solution
Sodium alginate is dissolved in the flavoured liquid and dropped into a bath of calcium lactate.

2 Gel coating
The calcium ions diffuse towards the centre of the droplet even after it is removed from the calcium bath.

3 Calcium lactate bath
Calcium ions displace the sodium from the alginate molecules and hold them together to form a gel.

Reverse spherification

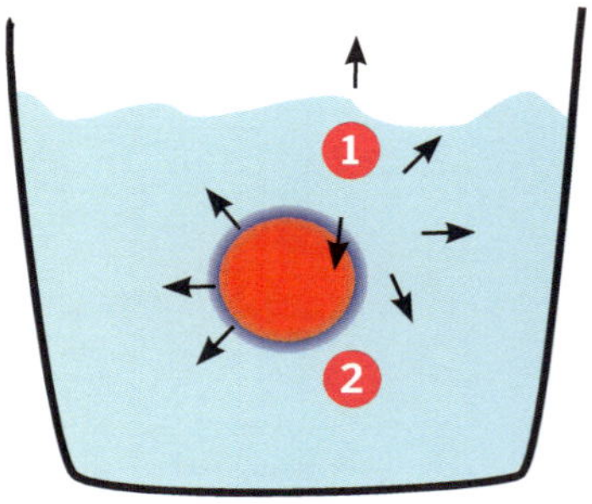

1 Calcium lactate solution
Flavoured liquid is infused with calcium lactate ar then dropped into a bath of sodium alginate.

2 Sodium alginate bath
Calcium ions diffuse from the droplet into the alginate, forming an outer gel layer until it is removed from the bath.

Sous vide

o ensure their food is cooked to perfection, many top hefs use a method called sous vide, which is French or 'under vacuum'. It involves sealing raw ingredients a vacuumed pouches and then placing them in a ater bath heated to a precise temperature. This ooks the food evenly from edge to edge, naintaining moisture and tenderness in the centre hile preventing the outsides from being overdone.

Despite the name, vacuum sealing isn't actually vital for sous vide, as temperature control is more important. However, it does have certain benefits. Sealing the food allows for more efficient heat transfer from the water and retains moisture and flavour during cooking. It also inhibits off-flavours that can occur when food is exposed to oxygen in the air. It's a process thought to date back to 1799.

Sous vide can make thick cuts of meat tender yet still medium-rare all the way through

"Vacuum sealing isn't actually vital for sous vide, as temperature control is more important"

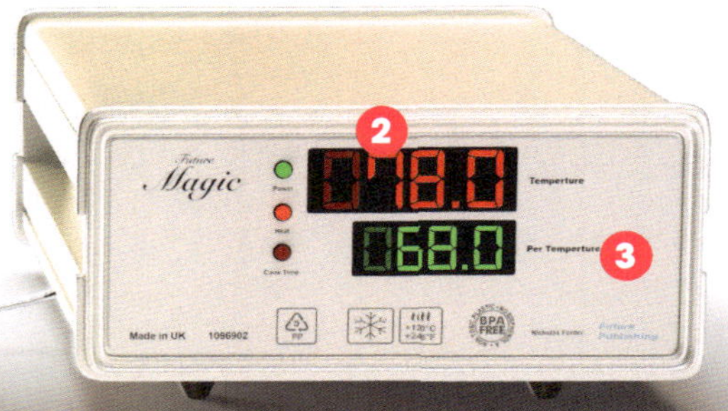

Efficient cooking Water transfers heat more ficiently than air, which eans food can be cooked lower temperatures.

2 Succulent flavour Cooking at low temperatures means the cell walls in food do not burst, providing a higher level of succulence.

3 Perfect 'doneness' Precise temperature control ensures proteins, such as fish, chicken and eggs, are cooked to perfection.

4 Safe to eat Cooking at exact temperatures and times ensures the food is safe to eat and determines expected shelf life.

5 Ideal texture When cooking vegetables, low temperatures ensure they are cooked thoroughly but maintain a crisp texture.

Foams

Thanks to science, chefs are able to present flavours in all sorts of interesting textures, including that of foam. To create an edible foam garnish, a surfactant such as egg white, gelatine, lecithin from egg yolk or agar-agar from seaweed is added to a flavoured liquid, and then air bubbles are introduced by shaking, blending or injecting nitrous oxide into the mixture. The surfactant contains both water-attracting regions that dissolve in the liquid and water-repelling regions, which stick to the air bubbles. This helps to reduce the surface tension of the air bubbles, stopping them from popping or escaping so that foam is formed.

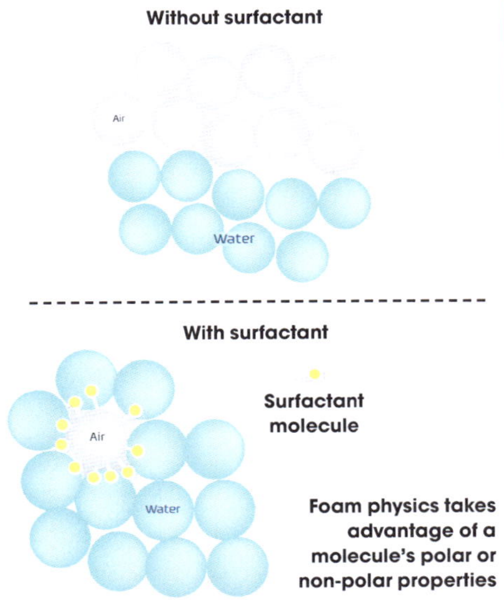

Foam physics takes advantage of a molecule's polar or non-polar properties

Emulsions

Oil and water don't naturally mix, so when a chef wants to create a deliciously creamy sauce in the flavour of their choice, they need the help of a special ingredient. Surfactants, such as egg whites and lecithin from egg yolks, enable the two liquids to mix in the same way they help keep the air bubbles inside edible foam. One part of the surfactant is attracted to the water, while the other is attracted to the oil, so it is able to bind the two together to form an emulsion.

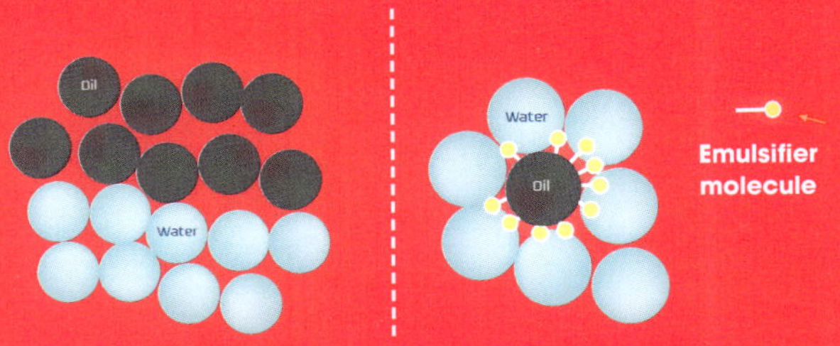

Cosmetic chemistry

Lotions and potions are packed with chemical science, but do they live up to the hype?

Mattifying makeup

Mattifying makeup products attempt to minimise shine by including ingredients that absorb oil and water. Clays are made from oxygen or oxygen and hydrogen (hydroxyl groups) and arranged into crystals with four or eight sides. At the centre of each is an atom of silicon or aluminium. These structures form sheets, and it's between these sheets that oil and water become trapped. Mattifying agents can be found in a variety of products, including powders, foundations, lotions, balms, gels and sprays.

Silicone elastomers are a synthetic alternative. They are made from chains of silicone, carbon, hydrogen and oxygen strung together to form branching webs. These expand when liquids are added, helping to lock moisture away.

Anti-wrinkle creams

Proteins are the major building blocks of the human body and collagen and elastin are two critical types found in skin. They form a web-like scaffold that holds skin cells in place; collagen provides structure and elastin provides springiness. However, as we get older, we produce less and less of both, and skin starts to lose its firm, flexible texture.

One solution employed by cosmetics giants is to add fragments of protein to their creams. This helps to smooth out wrinkles, but probably not in the way you were expecting. Rather than repairing the collagen and elastin scaffolding, the fragments work by improving skin texture from the outside. They go on damp and flexible, and as they dry out they tighten up. This tugs on the skin beneath, temporarily smoothing out the wrinkles.

> "Exfoliating washes contain fragments of sugar, nut shells or plastic that act as abrasives"

Whitening toothpaste

Whitening toothpastes work like a rough polish, using abrasive grains to scrape away the film of bacteria and pigments that continually builds on the surface of the enamel. However, they can't change the colour of the teeth beneath. For teeth whiter than a natural shade, chemical bleaching is the only option. Whitening strips contain carbamide peroxide, which breaks down chromophores, the parts of molecules responsible for their colour.

Waterproof mascara

Mascara colours the eyelashes using pigments like carbon black (made by the incomplete burning of petrochemicals) and iron oxides (which come in varying shades of red and brown). For ease of application, they are suspended in oils, waxes and water, forming a paste that can be spread onto the lashes with a brush. These carriers include beeswax, shellac (a type of resin made by lac bugs), lanolin (from sheep) and paraffin.

Waterproof mascaras tend to contain more waxes and oils than their water-soluble counterparts, helping them to resist moisture and stay on the lashes longer. But, because they've been designed not to dissolve in water, they can be a challenge to wash off. Oily makeup removers help to dissolve the waxy carriers that stick the pigments to the lashes. The mascaras themselves sometimes contain lubricants, like glyceryl stearate, which help the mixture to stay slippy.

Exfoliators

The skin is in a constant state of renewal, and as new skin cells are made, the old ones flake away. Exfoliators help to rub these from the surface, making the skin look smoother and brighter.

The textured surface of a washcloth is enough to gently scrub away some of the surface cells, but many treatments offer a deeper cleanse. Exfoliating washes contain tiny fragments of sugar, crushed nut shells or plastic that act as abrasives. Chemical peels use acids (commonly salicylic or lactic acid) to create a controlled burn, and microdermabrasion uses a rough rotating brush to scrape away even more of the skin's surface.

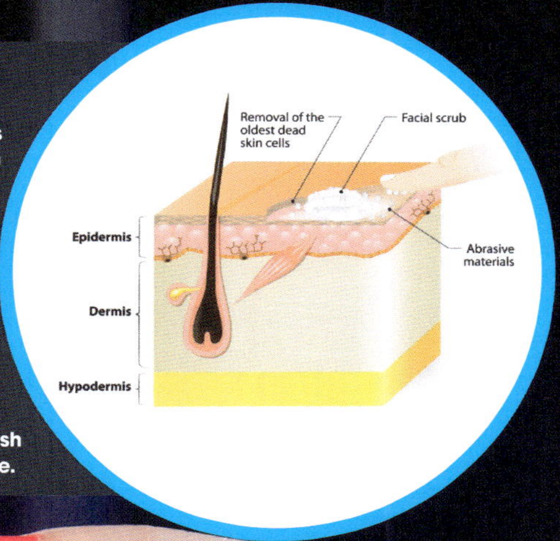

Shimmery shadows

The glitter and sheen in eyeshadows and highlighters is most commonly mica or bismuth oxychloride. Mica bends the light as it hits, while bismuth oxychloride has a pearlescent appearance. If mica is combined with titanium dioxide the way it reflects light changes, creating iridescent shades.

Makeup primers

Primers contain lots of ingredients that aim to keep makeup looking fresh all day. Silicones help to absorb moisture and oils, while waxes and polymers form a bridge that sticks cosmetics to the skin. Spherical silicone molecules coated with titanium dioxide help to diffuse light, evening out blemishes and creating an 'airbrushed' look.

Cosmetic myths

1 Bypassing the barrier

Creams might contain vitamins, proteins or even DNA and be accompanied by claims that the molecules will repair the tissue underneath. But skin is a barrier; not everything will be able to get through.

2 Stretch mark miracles

Hydrated skin hides blemishes better than dry skin, but stretch marks are scars, so no amount of moisturiser is truly going to remove them.

3 Hot water opens pores

Pores don't open and close, and there is absolutely nothing you can do to change their size. However, warm water can help to loosen anything that might be stuck inside your pores.

4 'Clinically proven'

You should always be wary of cosmetic science claims. Studies can be small or involve experiments using raw ingredients in test tubes, not the actual product on real people.

Shampoo science

What chemical processes occur within the foaming lather of shampoo?

Leaving hair unwashed for multiple days can cause it to feel heavy and greasy. This is because the scalp is continuously producing a natural oil called sebum. Without this sebum shield, the proteins within our hair would become damaged more easily. When there is a build-up of this oil and our hair is in need of a wash, water alone can't remove it very effectively.

Sebum largely resists mixing with water, especially if the water is cool, but when the water combines with shampoo it targets the oil. Molecules in shampoo, called surfactants, work to produce a frothy lather. They have two polar ends – one being hydrophobic (repelled by the water) and one being hydrophilic (attracted to the water). When rubbed into hair with both water and oil, the grease and dirt is carried off hair and down the drain.

Different qualities for all hair types

Anti-frizz shampoo often contains added silicones to make hair smoother. To achieve this, silicones act as a barrier against moisture. This stops hair from drying out, as well as preventing humid weather contributing to frizziness. Shampoos with more silicones aren't recommended for fine hair, as the weight they add can cause strands to break. Medical products like anti-dandruff shampoo contain ingredients to soothe discomfort experienced on the scalp and target more specific scalp issues. For dandruff sufferers, pyrithione zinc can be added to kill the Malassezia yeast that causes the condition.

Hairdressers use different shampoo types based on a customer's needs

What do you put on your hair?

These are the basic ingredients of shampoo

10-20%
Surfactants
These ingredients lower the surface tension between oils and water, allowing the shampoo to bind with grease and remove it during washing.

0.3-0.5%
Fragrance
Multiple ingredients can be added to create the smell of your shampoo. These may include natural chemicals such as menthol, or synthetically produced fragrances such as jasmine-smelling amyl cinnamal.

1%
Preservatives
Ingredients such as sodium benzoate, potassium sorbate or sorbic acid prevent bacteria from growing in the bottle.

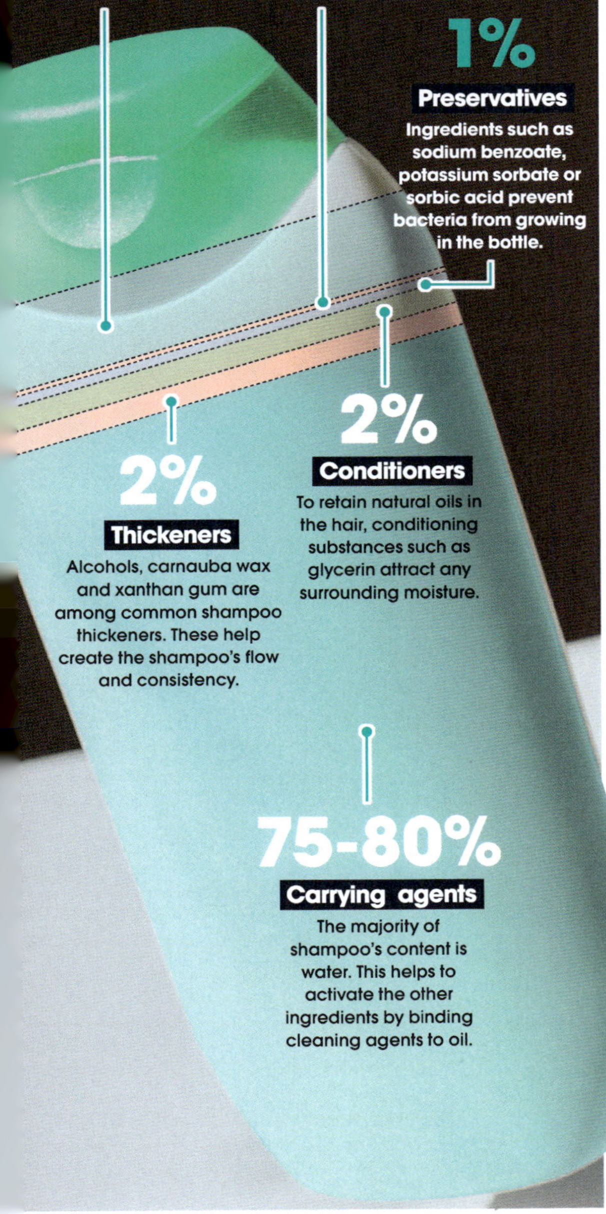

2%
Thickeners
Alcohols, carnauba wax and xanthan gum are among common shampoo thickeners. These help create the shampoo's flow and consistency.

2%
Conditioners
To retain natural oils in the hair, conditioning substances such as glycerin attract any surrounding moisture.

75-80%
Carrying agents
The majority of shampoo's content is water. This helps to activate the other ingredients by binding cleaning agents to oil.

The cleaning process

Take a close-up look at the chemistry of a hair wash

Oily build-up
The scalp produces a waxy substance called sebum; this stops hair from drying out but also attracts dirt.

Isolated conditioners
Conditioner molecules are positively charged, so they attract a fatty acid coating. This stops them from binding with the shampoo molecules' negative charges instead of the hair grease.

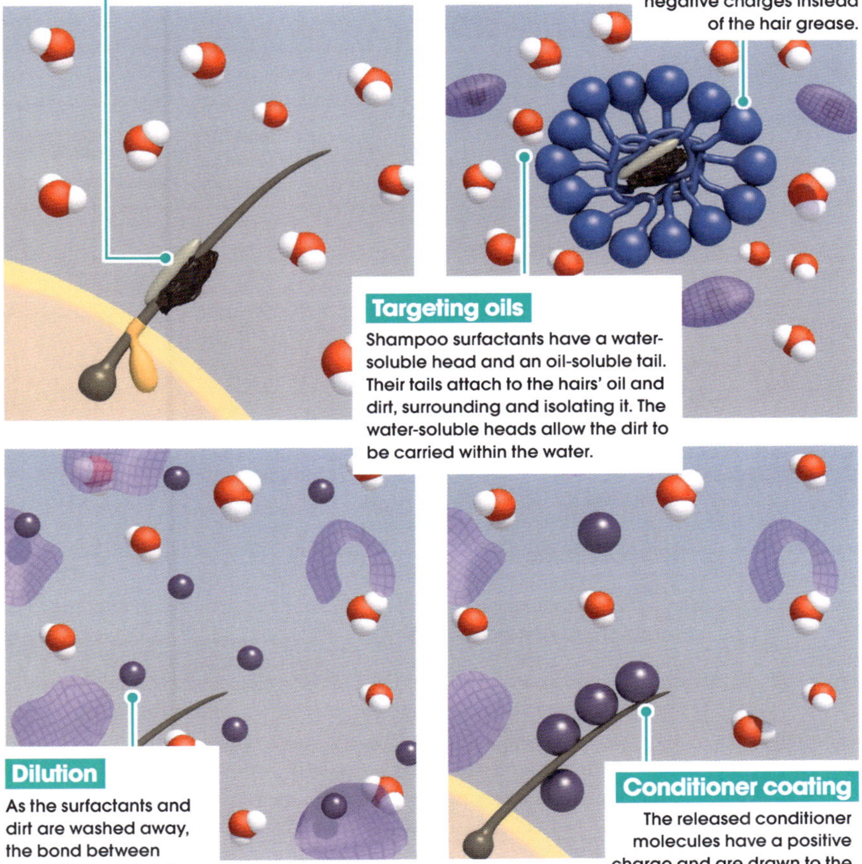

Targeting oils
Shampoo surfactants have a water-soluble head and an oil-soluble tail. Their tails attach to the hairs' oil and dirt, surrounding and isolating it. The water-soluble heads allow the dirt to be carried within the water.

Dilution
As the surfactants and dirt are washed away, the bond between conditioner molecules and their fatty acid coating weakens.

Conditioner coating
The released conditioner molecules have a positive charge and are drawn to the clean hair. This forms a moisturising coating.

Overwashing hair can make it too dry. Sebum glands can respond by producing extra oil and making hair greasier!

A violent criminal may have cleaned the area, but luminol still finds traces of blood

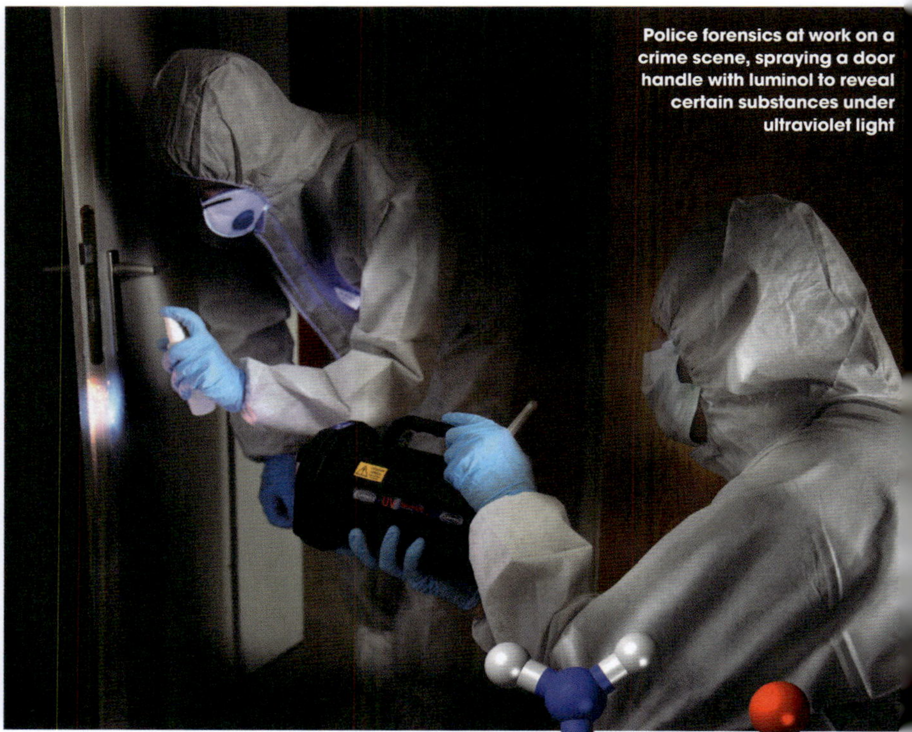

Police forensics at work on a crime scene, spraying a door handle with luminol to reveal certain substances under ultraviolet light

How luminol works

The blue glow of this forensic chemical can indicate a violent crime has taken place

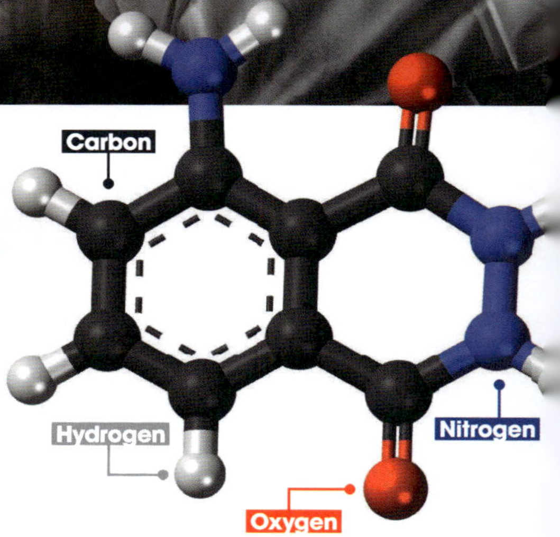

Police may well believe that a serious crime has taken place, but without any obvious signs of violence, their investigation could come to a grinding halt. That's when the forensic scientists step in to look for seemingly invisible clues. Criminals often attempt to clean up a crime scene by mopping up any spilt blood, but just because you can no longer see it doesn't mean it isn't there, and a quick spray of luminol can often reveal the gory details.

Forensic scientists began using the luminol test in 1942, having discovered that when the chemical compound came into contact with an oxidising agent, such as blood, it caused luminescence. The eerie blue glow that is produced highlights exactly where blood has been spilt, and often lights up other vital clues through the blood stain patterns, such as how the crime was committed, if a body was moved and how the perpetrator left the crime scene. Since tiny particles of blood can cling to surfaces for many years, even older crime scenes can benefit from the luminol test.

The science behind this amazing forensic investigator's tool is very similar to that of a simple glow stick. Luminol powder, made up of carbon, hydrogen, oxygen and nitrogen, is mixed with a number of other liquid chemicals that react with the iron in haemoglobin, instigating a process known as 'chemiluminescence'. When the room is plunged in darkness, the telltale blue glow reveals any bloodshed.

Although luminol is a useful forensic aid, it isn't always helpful. Substances other than blood can cause the glow, such as the sodium chlorate in bleach or certain enzymes in faece It is therefore vital that any chemiluminescenc that occurs is followed up with another test to make sure blood has been found.

The Huber murder case

In 1994, John Joseph Famalaro was taken in for questioning after a body was discovered in his home. Police investigated his previous addresses, including a warehouse he once owned in Orange County. The place appeared clean, but when luminol was sprayed across the walls, it lit up. The warehouse was covered in blood. After DNA samples had been removed and analysed, it was confirmed that the blood belonged to Denise Huber, who had gone missing three years earlier.

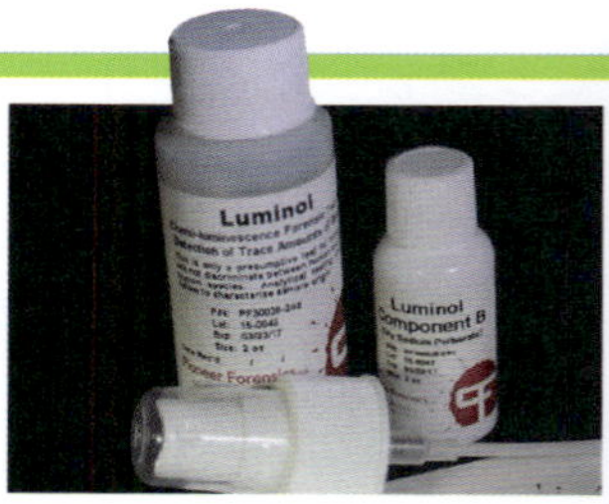

A bottle of luminol spray, like that used in the case

Did you know?
German chemist H O Albrecht first discovered luminol's glow in 1928

The science of chemiluminescence

It looks like a ghoulish magic trick, but how does it work?

The oxidising agent
A strong oxidising agent such as hydrogen eroxide is added to the luminol, also known as amino-2,3-dihydro-1,4-phthalazinedione.

2 The basic solution
A basic liquid solution is made by adding an alkali – sodium hydroxide – to the luminol, which can now be poured into a spray bottle.

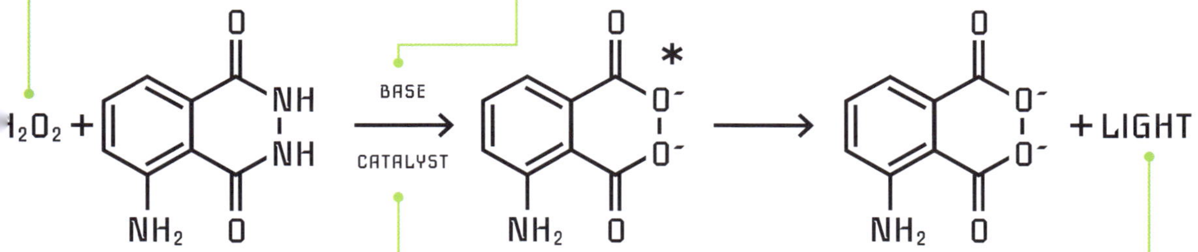

3 Detecting the catalyst
A catalyst is required to accelerate the action between the luminol and hydrogen eroxide, such as the iron found in haemoglobin.

4 The blue glow
Luminol loses nitrogen and hydrogen during the reaction and gains oxygen atoms, which releases energy. Any excess energy appears as photons of light.

Luminol triggers

Blood **Bleach** **Faeces** **Urine** **Horseradish**

Luminescence in nature

Some living things are capable of producing their own light, known as bioluminescence. The angler fish lures its prey by dangling a glowing line, enticing smaller fish to come closer. Other predators of the deep stun their prey by emitting a sudden flash of light, confusing them. The pulsing lights of insects signal they're ready to mate, while some mushrooms glow to encourage moths to land on them and spread their spores. Like luminol, bioluminescence occurs through a chemical reaction. However, unlike the forensic investigative tool, bioluminescence requires the chemical luciferin.

5 Facts about luminol

1 Blink and you'll miss it
The chemiluminescent glow will not last very long after the luminol has been applied. In fact, it starts to fade after approximately 30 seconds.

2 1:1,000,000
Luminol is so sensitive that it can identify the presence of blood at a dilution of up to one part in a million, which means even the tiniest speck can be detected.

3 Hold the sauce
Unfortunately for forensic investigators working in a kitchen crime scene, luminol also begins to glow when it comes into contact with the peroxidase enzymes found in horseradish.

4 Confirmed by a second test
Because the luminous reaction is something that can occur with a number of other substances, forensic scientists may also carry out a phenolphthalein test to confirm the presence of blood.

5 Imitating nature
The strange glowing phenomenon that occurs when luminol is sprayed onto a bloodstain can also be seen in the insect world with glow worms and fireflies.

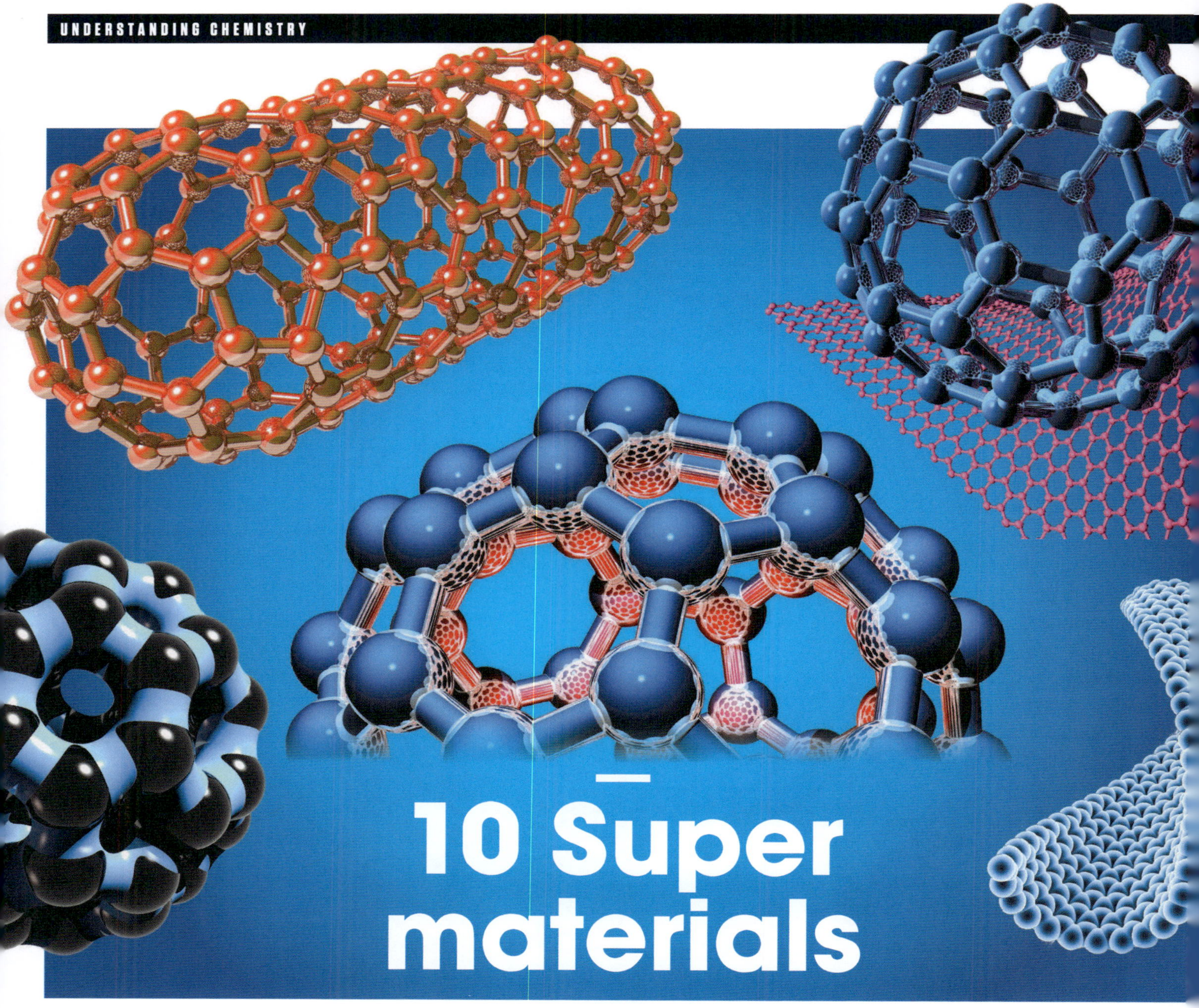

10 Super materials

How are we enhancing Mother Nature's design to develop the new-and-improved materials tomorrow's world will be made of?

With natural resources dwindling and some simply no longer meeting our needs, a new range of 'super materials' are now being developed in labs around the world. Designed to increase efficiency, these substances are new compounds that build upon and improve what's currently available, to be the best in a particular field.

Natural materials have been used for decades and even centuries to perform many of our day-to-day tasks, from conducting electricity to insulating heat, but super materials take things to a whole new level. The emphasis is now geared towards the best and only the best. Nothing less than total conduction, extreme strength or complete insulation will do. Essentially these materials will do the job far better than anything that has gone before.

Whether based on a natural substance or an improvement on a manufactured effort, super materials will become increasingly important in a world searching for sustainable energy sources. However, many questions remain. Can we harness these materials and mass-produce them? Will they be available to the general public? Are they as good as they seem?

Over the following pages we present our pick of ten of the most impressive super materials that look set to reshape our future.

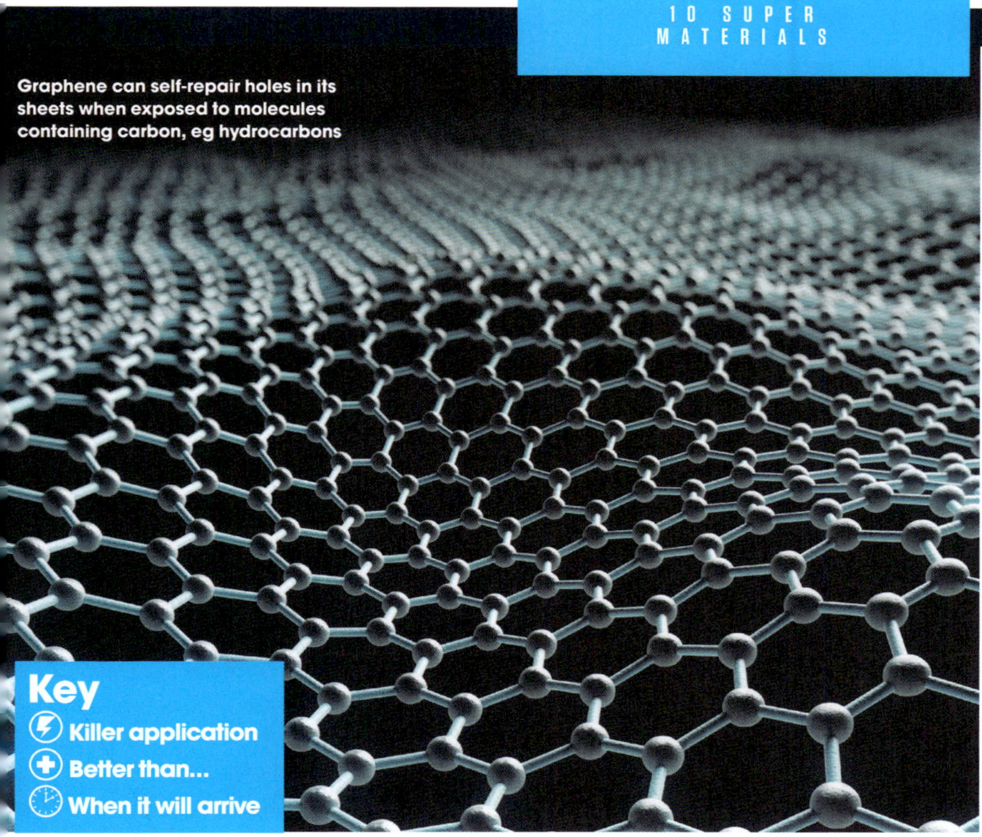

Graphene can self-repair holes in its sheets when exposed to molecules containing carbon, eg hydrocarbons

Key
- Killer application
- Better than...
- When it will arrive

1 grand piano

A strand of graphene as fine as a pencil point is able to hold up a 450kg piano

Graphene Stronger than diamond

: super materials had a star, graphene would e it. Composed of a single layer of graphite arbon atoms in a honeycomb pattern, its tructure is stronger than diamond. It was first reorised as far back as the mid-20th century ut only gained recognition for its astonishing roperties when Andre Geim and Konstantin 'ovoselov experimented with it and went on to laim a Nobel Prize in 2010.

In addition to strength, graphene is also amed for its excellent conduction of both eat and electricity. Atomic force microscopy as proved it is, at the very least, 100 times ronger than steel and can be stretched by up 20 per cent of its own size. It has been used or all manner of things, such as a coating naterial to nullify lightning strikes, increasing nergy storage in batteries and making ouchscreens more responsive.

Its coating properties in particular help stop corrosion and prevent micro-organisms from spreading. The electrons within it travel at a hundredth the speed of light, as if they carry no mass. Graphene's tiny size makes it ideal for small electronic devices, as its high thermal conductivity enable them to dissipate heat while still maintaining power.

Graphene is actually the source of many other super materials and is the parent form of carbon nanotubes and buckyballs. However, it was only experimentally isolated on its own accord in the 2000s by the aforementioned Nobel Prize winners.

There are currently only a few ways of producing graphene: mechanical or thermal exfoliation, chemical vapour deposition and epitaxial growth. None of these methods are exactly geared for production on a large scale, so a new way of creating the super material has been proposed. This involves oxidising the graphene to turn it into graphene oxide, which is easier to contain and transport. However, this method is still in its early stages.

Adding this simple carbon allotrope to a variety of surfaces and devices is surely the future as the human race looks to establish ever-more efficient materials. Some are sceptical of the potential of this substance and it's admittedly hard to believe that one material can have so many impressive properties, but graphene undoubtedly still has much to offer.

- Counteracting lightning strikes
- Copper wires at conducting electricity
- Already around, with its uses increasing

Five ways we can use graphene

1 Lightning catcher
Outstanding electrical conductivity means it can not only nullify lightning, but perhaps even harness it. This has never been done, but graphene could be the material for the job.

2 Wires
Copper wires are found in all electrical circuits, but graphene can transport 1,000 times the density of copper. Increasing electrical efficiency would cut emissions and fuel use.

3 Coatings
Graphene is very useful when it comes to creating composite materials and can also coat plastics or metals to improve their properties, such as electrical conduction or making them stronger.

4 Touchscreens
Handy for touchscreens on smartphones and tablets, graphene is transparent and can transmit 97.7 per cent of light. Its strength also sees off scratches while still keeping that vital flexibility.

5 Bio-engineering
Experts suggest the material could have the capability to monitor glucose and cholesterol levels. What's more, it is thought it can aid tissue rejuvenation and even cancer treatments.

Another type of natural plastic called coleoptera is made of beetle shells; inset is an example of shrilk

Shrilk

The next generation of plastics

Composed of silk proteins and shrimp shell, shrilk combines biodegradability with excellent flexibility and strength. Based on similar substances in the animal kingdom, shrilk's roots lie in the material found in shells and insect wings. The hope is that the material can eventually replace plastic, which would lessen the impact and size of landfill sites. Like plastic, it's inexpensive and can be used to make clothing, bags and many other everyday products.

In addition to shrilk, there has also been progress with another plastic formed from dead beetle shells – known as coleoptera. This contains chitin, a natural polymer that boasts the light weight and flexibility of conventional plastics but breaks down far more easily.

Shrilk and other biodegradable plastics are great examples of where the fields of biology and engineering create the ultimate material solutions.

- Potential to replace plastic
- Lighter than aluminium and equally strong
- More research needed to be mass-produced

Iron nitride

Strongest magnet in the world

The most magnetic material on Earth is a mix of iron and nitrogen with the chemical symbol Fe16N2. It's claimed that the material has the strongest saturation magnetic flux density of any human-made substance. This means that the strength of magnetism is the most per unit of molecule within the iron nitride, making it hugely magnetic over its entire surface. Powered by ferromagnetism, iron nitride is electrically uncharged but that doesn't affect its power.

Every electron within the material acts like a tiny magnet. The Fe-N clusters increase electron contact, which intensifies the charge. It's so magnetic that it exceeds the predicted limit of magnetism for a single material. Iron nitride has taken over from the previous holders, neodymium and iron cobalt, to claim the plaudits for the most magnetic material on Earth. It's at least twice as magnetic as its rivals (rated at 130 megagauss oersteds).

Magnets of this magnitude are used in industry and engineering to increase the efficiency of producing power. They are often used as electromagnets within transformers, for example.

- Making disk drives more effective
- Neodymium and iron cobalt
- Still in the testing stages – watch this space for more

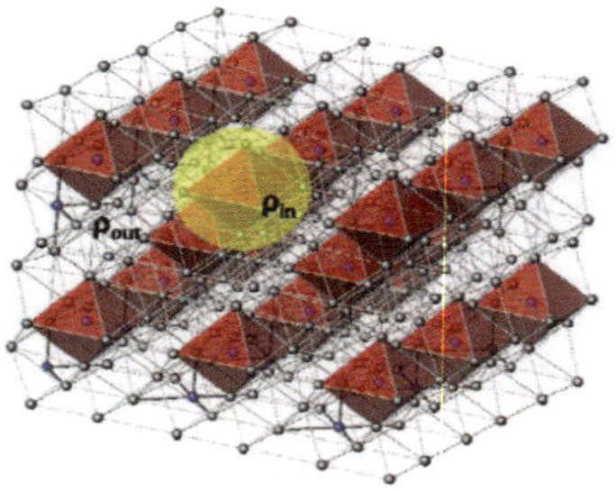

"High magnetic fields play a critical role in developing new materials that affect nearly every modern technology. The vast scope of work currently underway includes the study of new superconductors with the potential to revolutionise how power is stored and delivered. There's also a search for new medicines and analysis of petroleum samples that could lead to better oil extraction."

Greg Boebinger,
Director at the National High Magnetic Field laboratory, Florida State University, Florida, USA

Stanene

The perfect conductor

Composed of a single layer of tin atoms, stanene is said to be able to conduct electricity with 100 per cent efficiency at room temperature.

This astonishing electrical conductivity has seen it dubbed by some as the new graphene and it's expected to play a big role in the future of computer chips.

Stanene is part of a group of topological insulators that conduct electricity only on their outer edges. Being only an atom thick, the electrons can travel through the material with no resistance, which improves efficiency. However, stanene is the first of these topological insulators to work at room temperature. Difficult to produce due to its minuscule size, the material is still in the developmental stages. There's also talk of adding in a layer of fluorine to increase conduction efficiency at higher temperatures as well.

- Boosting the processing speed of computer chips
- Potentially even graphene
- Far from wide-scale production

The hydrophobic qualities of plants like nasturtiums have been applied to glasses in traffic control units, medical slides and circuit boards (inset)

Super-hydrophobic materials

Completely waterproof

Human-made waterproof materials have paled in comparison with natural examples such as the lotus leaf and insect wings – until now. Known as the most waterproof material ever, super-hydrophobic surfaces have been developed at the Massachusetts Institute of Technology (MIT) and are inspired by butterfly wings and nasturtium leaves.

Often referred to as the lotus effect, nature's waterproof materials defend themselves from water through a special structure. They are covered by bumps or hairs that when exposed to liquid can direct it away from the body. Various manufactured materials take advantage of this technique, including synthetic silicon, polymer microposts and electro-deposited copper. These coatings, like the organic inspiration, enable water droplets to bounce off a surface to keep it dry. The materials have small ridges that break up the water on the surface and disperse it before it can soak through.

Some of these materials are being pushed even further, with efforts to make them repel ice and snow too. Hydrophobic materials are perfect for everything from clothing to tents and vehicle coatings.

- A de-icer that will rapidly clear snow and ice
- Lotus leaves have finally been surpassed
- Already available in clothing and more

"To be super-hydrophobic, a material requires both hydrophobic chemistry and roughness. The trapped layer of air, under certain situations, may act to reduce the drag on an object passing through water."

Michael Newton,
Nottingham Trent University, UK

Buckyballs

he original super material

fectionately known as buckyballs, buckminsterfullerene is one of eight rbon allotropes that include diamond, graphene and carbon anotubes. Discovered by accident, buckminsterfullerene can be nsidered the daddy of super materials. Its discovery paved the way r the modern era of nanotechnology and proved without a doubt that aterials with extreme properties could be found and worked on. It has nce led to the discovery of carbon nanotubes, as scientists were couraged to enhance carbon allotropes further in the search for the ext carbon nanomaterial.

The allotrope is shaped a bit like a football, with a hexagonal and mmetrical polyhedral structure. A tough skeleton of 60 carbon atoms akes buckminsterfullerene even stronger than diamond under certain onditions. Every carbon atom within this super material also has three nds, resulting in its incredible strength. Uses include photovoltaic plications in solar panels and the inhibition of a protein in the HIV us to stop it replicating. Some have even said it could limit oxidative ess of cells in the body that cause ageing.

- Could combat HIV and cancer
- Diamond when tested for hardness under high pressure
- Widely used since 1985

"Superglue is a polymer-based adhesive of the cyanoacrylate type that is polymerised upon contact with a surface and moisture. Some of the uses of these adhesives can be in the form of coatings, fillers, forensics, or even for medical uses like closing (open) wounds."

Rigoberto Advincula,
Adjunct professor of macromolecular science and engineering at Case Western Reserve University, Ohio, USA

Molecular superglue

Most adhesive glue

First there was glue, then there was superglue, but now there is molecular superglue. From the cyanoacrylate (instant glue) family of adhesives, it is designed to be the most adhesive material on the planet and will primarily be used to fight disease.

Genetically engineered from proteins, the glue is polymer-based and formed from nanotechnology that bonds molecules together to form tough covalent bonds. The technology enables the protein to react with itself to form a tight lock. Most effective when used thinly, the glue is made from the proteins of the streptococcus pyogenes bacteria, enabling it to hook on to human cells.

Even more impressive, the glue can be designed to be selective to what it sticks to. This is essential, as an adhesive this strong would cause havoc if it got stuck to the wrong objects!

- Closing up wounds in seconds
- Any previous superglue – and a whole other league to PVA glue
- Here now, but its uses are not fully confirmed

Solids that are lighter than air

Created by removing liquid from a gel, aerogels are the world's lightest solid materials. High in strength and low in density, they are mesoporous, which means they contain lots of tiny pores, contributing to their low density. There are various types of aerogels, all with different functions and abilities.

First, silica aerogels have an extremely low thermal conductivity and can be used as super-insulators. The most common of the gels, these have even been used on expeditions to Mars.

Carbon aerogels can store high amounts of energy and are ideal as fast-charging super-capacitors. They are predicted to be used as a new type of faster charging and discharging battery in mobile phones and electric cars.

But that's not all – metal aerogels combine the properties of the two substances. Being highly conductive and having a high surface area, X-ray optics and hydrogen storage are just two of the possible functions for this hybrid material.

- Protection in firefighter suits
- Surpasses all other heat insulation
- Currently used, but development continues

Titaniumfoam
The metal that can replace bone

Metal foams are generally solid metals filled with tiny holes, known as 'cells', and up to 95 per cent of their volume can be air. Their biggest selling point is that while they are light and porous, they retain much of their original strength. Made from a mix of metal powder and polyurethane, a binding agent fixes the two substances together under heat.

Titanium foam in particular is tough but at the same time has very similar properties to bone. Experts predict that bone will be able to naturally regrow around it, making this material a very attractive prospect for mending breaks and fractures. Also corrosion-resistant, it can endure nearly all chemicals, making it useful not just as a biocompatible material in the body but also for aerospace components.

- Currently used on drones and lightweight aircraft
- Many metals used in construction
- Expected to be used in bone reconstruction

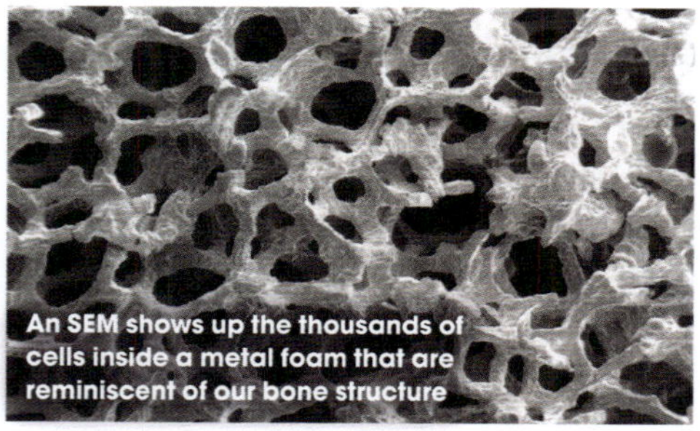
An SEM shows up the thousands of cells inside a metal foam that are reminiscent of our bone structure

"Titanium is light, strong and, most importantly, corrosion-resistant. The most vital application of titanium foam is as artificial bones, because it can be tailored to have similar mechanical properties to human bones and the porous structure is conducive to ingrowth of tissue cells."

Yuyuan Zhao,
Head of the Centre for Materials and Structures, University of Liverpool, UK

Aerogels Make your own diamond

Famed for their beauty and toughness in the natural world, diamonds are becoming increasingly rare. Steve Coe, formerly of synthetic diamond manufacturer Element Six, takes us through two ways they're re-creating these gems for drilling, optics, acoustics and more.

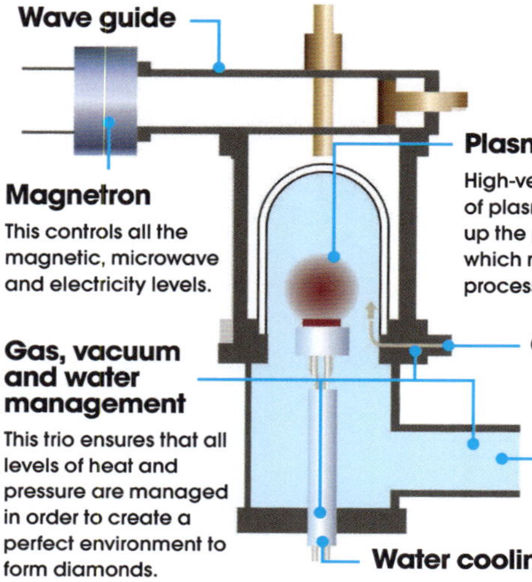

Chemical vapour deposition (CVD)

Chemical vapour deposition uses a hydrocarbon gas mixture, where the diamond is produced in a vacuum system below atmospheric pressure, with carbon atoms supplied from a gas such as methane and deposited in layers onto a substrate.

By passing microwaves through the gas to generate a plasma, at temperatures of around 2,000°C, atomic hydrogen is created. This enables impurities in the form of graphite to ensure that only the diamond carbon is deposited.

This technique enables tightly controlled growth conditions, eliminating impurities and enabling the engineering of various properties into the diamond material.

High-pressure high-temperature diamond synthesis (HPHT)

This is a synthesis process by which synthetic diamond is created under enormous pressure and temperatures to replicate the Earth's natural process.

The proprietary belt-press technology contains two large anvils to apply hydraulic force to a capsule at the centre.

This capsule contains graphite and a metal catalyst, which react during the process to form diamond. The 15,000 atmospheres of pressure applied to the capsule is the equivalent of taking the Eiffel Tower, inverting it and placing it on a soda can, then turning the temperature up to 1,500°C – the melting point of steel.

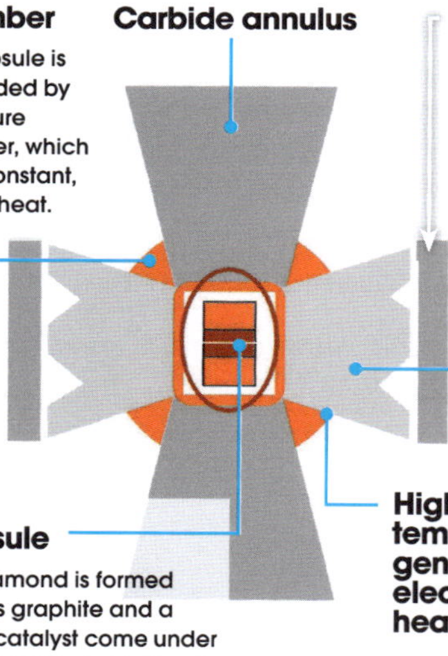

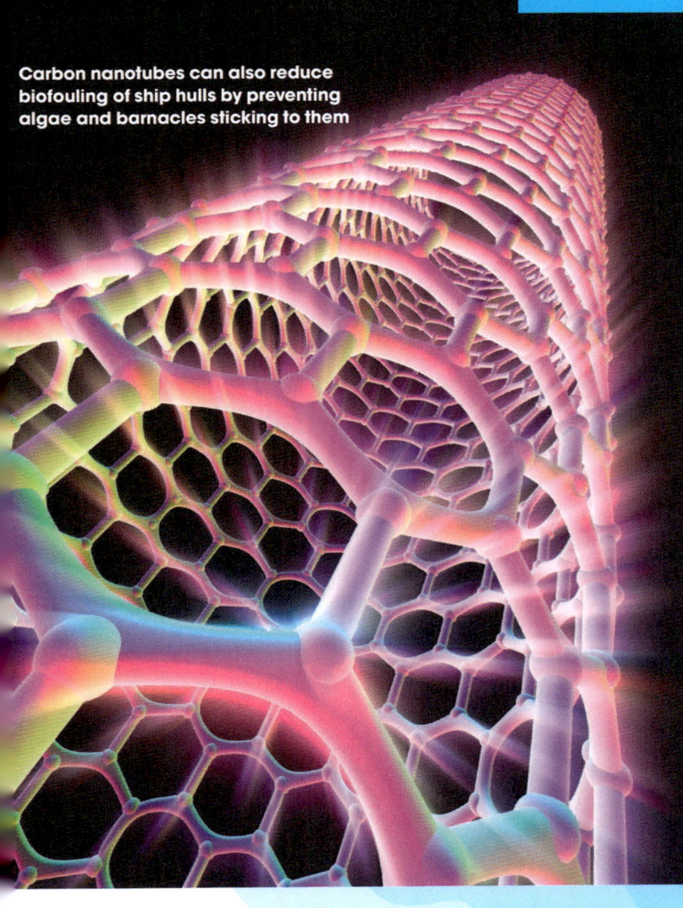

Carbon nanotubes can also reduce biofouling of ship hulls by preventing algae and barnacles sticking to them

Carbon nanotube

Absorbs over 99.5 per cent of light

Carbon nanotubes are said to outperform Kevlar and steel in strength. By having a high specific strength, carbon nanotubes can absorb high impacts by spreading out the force. As well as being strong, the fibre is also ductile and malleable. These characteristics render the material useful as a possible replacement for steel, as well as being applied as synthetic muscles and body or vehicle armour.

If all that wasn't enough, the computing world has also found a use for this super material with its potential to be a long-term replacement for today's silicon computer chips.

Moreover, certain types of carbon nanotube developed by NASA are said to be the darkest material known to humans, due to their ability to absorb over 99.5 per cent of photons. This is particularly useful for stopping stray light interfering with sensitive equipment on probes and spacecraft. This property also offers huge potential for more effective solar panels.

- Absorbing multiple wavelengths of light
- Silicon transistors in electronic devices
- Environmental concerns have stalled its progress so far

800°C

A nanometre of carbon nanotubes is heat resistant enough to withstand lava

"Carbon nanotubes are molecular-scale tubes of graphitic carbon with outstanding properties. They are among the stiffest and strongest fibres known, with a breaking strain around 50 times higher than steel. Carbon nanotubes have an important advantage over graphene, in that they are stiff and strong in compression as well as tension; graphene can't withstand any compression."

Peter Harris, Reading University, UK, and author of *Carbon Nanotube Science*

5 Natural inspirations

1 Lotus leaf

When raindrops land on the leaves of the lotus, they cannot settle on the plant. This is down to microscopic bumps over the surface, which increase the contact angle. As the water hits these protrusions, with pockets of air trapped in between, it beads up into spheres and rolls off.

2 Gecko feet

Geckos have the ability to cling to surfaces with adhesive pads called setae on their feet. However, this skill comes undone when the weather is wet, hence why scientists today are looking to advance on this natural ability by making a waterproof adhesive.

3 Kawazulite

This mineral is a conductor on the outside and an insulator on the inside. It looks set to play a big role in future synthetic insulators.

4 Spider silk

Tough and adaptable, spider silk has been used for everything from fish nets to gun crosshairs. Humans have created similar synthetic products such as Kevlar, however these generate lots of pollution.

5 Gold nanoparticles

Also called nanogold, these tiny particles are 500 times smaller than the width of a human hair. They have excellent molecular-recognition properties and are able to detect the proteins on cancer cells by using specialised antibodies.

How we use oil

Known as 'black gold', oil's versatility has fuelled progress – and problems

Oil shapes the modern world, giving humans enormous power, but creating massive environmental problems too. It's strange to think, but much of the furniture we sit on and fill our homes with and everything else plastic that surrounds us owes its existence to oil. Likewise, burning oil and using its energy is still what lets most of us travel nearly anywhere we want.

How can oil – thick black, brown or yellow goop that someone pumped up from deep underground – be so useful? It's because it's rich in the element carbon. Carbon atoms readily bond to each other, and to atoms of other elements. Plugged together like tiny building blocks, these atoms make many different-sized chains and other shapes. Some help form cells in animals – including us humans – and plants.

Over millions of years, formerly living material sinks to the bottom of lakes and seas. It gets buried there and slowly changes into a valuable, sticky liquid. We pay an enormous industry billions a year to suck that oil out. Then companies separate it into different parts, each of which does a useful job. However, every barrel of oil we pump up brings both challenges and benefits. When we separate out the parts of oil and then make things from them, we usually want them to last a long time. So when we throw them away, they stay in our environment and pollute it. Perhaps even worse, when we burn oil for energy, its carbon atoms combine with oxygen in the air to become carbon dioxide. This acts like an invisible blanket in the atmosphere, trapping heat, leading to dangerous global warming. Oil's usefulness means we still use it, and will keep doing so. Yet we must find a way to stop oil's problems outweighing its benefits.

Because plastic is durable, it can stay in the ocean for a very long time

Cleaning up our oceans

"Industrial pollution and the discarding of plastic waste must be tackled for the sake of all life in the ocean," said naturalist David Attenborough in the 2017 documentary series *Blue Planet II*.

Companies have tried to recycle plastics to fight the plastic problem for a long time. However, very few of the many types of plastic are easily recyclable. In European Union countries in 2017, nearly a third of all plastic was recycled on average. In the US in the same year, less than a tenth of plastic generated was recycled.

Research might be able to help. Some scientists are developing new methods to make recycling easier and more profitable. Recently others have helped speed up how fast bacteria eat up plastic. That would help return the plastics to the oil-derived materials they were made from. But not wasting plastic products in the first place remains an important course of action.

Carbon's flexible backbones

Carbon can bond to four surrounding atoms, so can form chains and many other shapes

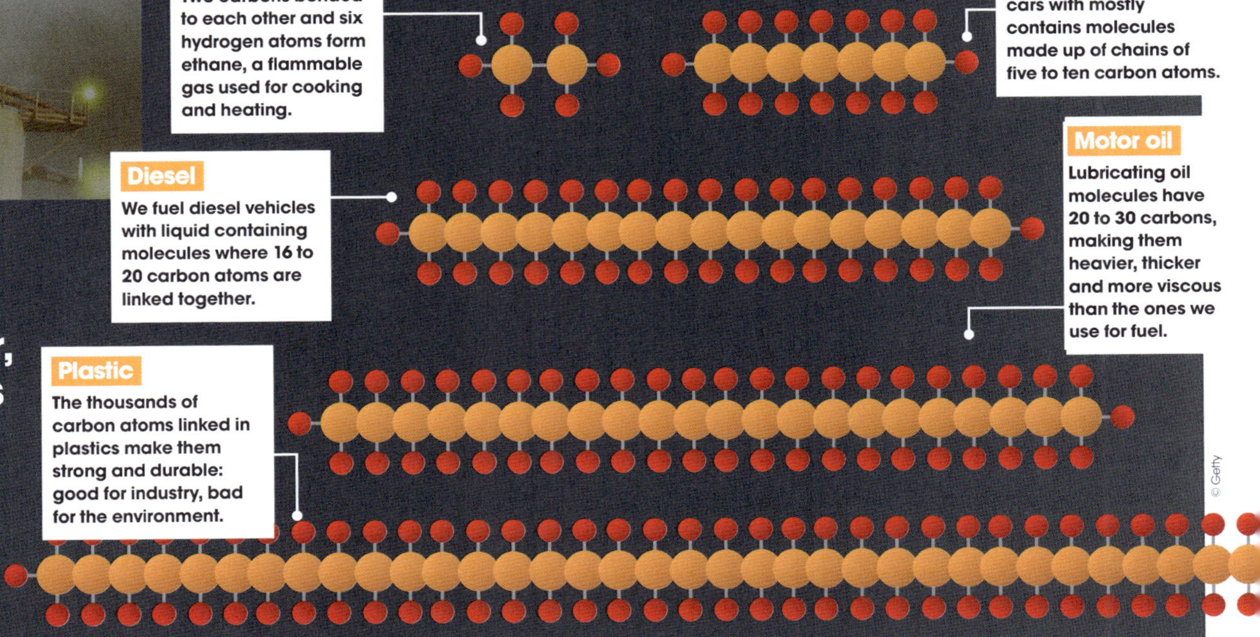

Ethane
Two carbons bonded to each other and six hydrogen atoms form ethane, a flammable gas used for cooking and heating.

Petrol
The liquid we fuel some cars with mostly contains molecules made up of chains of five to ten carbon atoms.

Diesel
We fuel diesel vehicles with liquid containing molecules where 16 to 20 carbon atoms are linked together.

Motor oil
Lubricating oil molecules have 20 to 30 carbons, making them heavier, thicker and more viscous than the ones we use for fuel.

Plastic
The thousands of carbon atoms linked in plastics make them strong and durable: good for industry, bad for the environment.

"Carbon atoms bond to each other, and to atoms of other elements"

A refined process

Cooled columns in refineries separate oil into different useful parts according to their boiling temperature

This valley used to be filled with ice, but climate change has melted it

Because plastic is easy to shape, it's used to create many items, like furniture

Crayons
Paraffin wax has around 20 to 40 carbon atoms per molecule. It is combined with different colour pigments to make crayons.

Wax paper
Applying petrochemical naphtha wax to paper makes it waterproof because water doesn't mix very well with oily substances.

Candles
In the heat of a flame, naphtha wax melts and then boils to form a gas that burns.

Virgin light naphtha
This contains most of the hydrocarbons with six or fewer carbon atoms.

Naphtha
Petrochemical naphtha wax has molecules with nine to 11 carbon atoms, sometimes looped into

Natural gas
The gas that many people use to cook and heat their homes includes the small hydrocarbon molecules methane and ethane.

Ethylene
Also known as ethene, small ethylene molecules can join up with each other to make the very common plastic polythene.

Refinery gases
These gases are by-products of oil refinement, but are also useful in their own ways.

Polypropylene
Some very small petrochemical gas molecules can connect with each other to make plastic. Propylene makes polypropylene, which is used in Tupperware.

<30°C

30–85°C

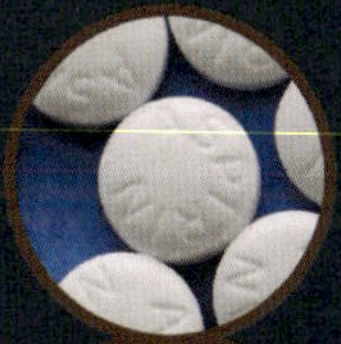

Aspirin
Many drugs originate from oil-based chemicals. For example, aspirin is made from phenol.

Hair dye
Phenol molecules are often highly coloured, meaning that they're used in hair dyes.

Sunscreen
Phenols can offer protection from the Sun, so are used in sunscreen.

Phenol
A volatile, crystalline solid, it is mildly acidic and can cause chemical burns, so must be handled with care, though it's used in many household products.

Petroleum blending components
These are the different grades of

Petrol
After separation from crude oil, the petrol that powers many cars is usually a clear, light-yellow colour.

65-175°C

175-235°C

235-345°C

345-565°C

>565°C

heavy naphtha
Contains most of the hydrocarbons with six or more carbon atoms.

Thermosets are irreversibly hardened by heat.

Thermoplastics
These materials become softer when heated, and harder when cooled.

Epoxy resin
Propylene from naphtha can go into making epoxy resins, which are thermoset.

Synthetic rubber
Once taken from trees, scientists found they could make it from naphtha.

Vinyl
The black records on which music is recorded are made from polyvinyl chloride (PVC) plastic.

Nylon
Breaking down naphtha into rings of six carbon atoms produces one of the ingredients that makes nylon.

Teflon
Teflon adds fluorine atoms to ethylene, which can come from either natural gas or naphtha.

Mineral oils
Viscous mineral oils are made up of long-chain carbon molecules arising from refining, and are often used as lubricants.

Motor oil
Lubricant liquids in vehicles and other large machines are made from molecules that have long chains of carbon atoms.

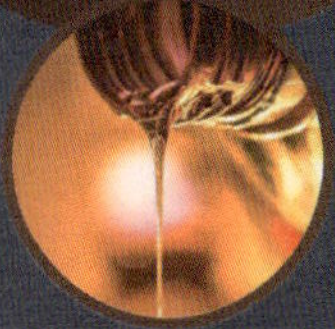

Petroleum jelly
Petroleum jelly's soft and slick texture comes from its mixture of hydrocarbons, mainly with more than 25 carbon atoms in them.

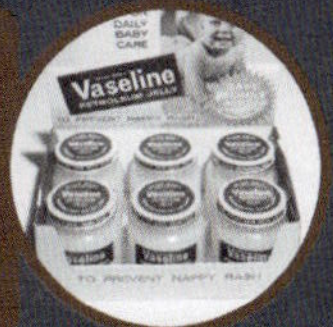

Moisturiser
Using petroleum jelly to form an oily barrier on your skin can stop it losing moisture.

Lip balm
The fact that oil and water don't mix means petroleum jelly can stop water escaping lips.

Crude oil
Refining crude oil by heating it allows its different components to be separated and removed for use for many purposes.

Bitumen
The thick, semi-solid bitumen left behind after oil refining can be used to make roads and roofs.

Kerosene
Also known as paraffin, this highly flammable derivative of petroleum is used as a fuel in aviation and households.

Camping fuel
The fact that kerosene contains a lot of energy makes it perfect for hikers who must carry small stoves.

Jet fuel
Aircraft are powered by kerosene because it burns well and contains a lot of energy.

Diesel
As cars that use diesel fuels pollute city air and can damage people's health, they are becoming less popular.

Diesel fuel
Diesel engines don't require a spark to ignite like petrol ones, using compression of the inlet air mixture instead.

Heating oil
Houses not connected to other heating networks are warmed by oil-powered heating, which is similar to diesel or kerosene.

Residual fuel oil
This is one of the cheapest oil products because it's made of very long carbon chains, making it very sticky.

Ship fuel
Big ships use very thick residual fuel oil, which needs to be heated up in order to be pumped into their engines.

Power plant fuels
Using residual fuel oil to generate electricity emits lots of planet-warming carbon dioxide, so this is generally becoming less common.

"Oil is useful because it's rich in the element carbon"

Understanding plastic

From toothbrushes to computer parts, meet the material that has it all

After ushering in a new era of affordable consumer goods in the Fifties, plastic has since conquered the world. Plastics are incredibly versatile. They can be hard or soft, opaque or transparent, flexible, stretchy or brittle, and much more. But all plastics share one defining property: they can be moulded into almost any shape.

Synthetic plastics are manufactured by joining monomers – small carbon-containing molecules – into long chains called polymers. Most industrial plastics are made from monomers extracted from oil or gas. Combining different types of monomers into a variety of arrangements creates plastics with a wide range of properties. Inside a plastic, polymer chains entwine like a bowl of spaghetti, making plastic extremely durable. On the flipside, plastic takes centuries to break down.

There are two main methods for joining monomers into polymers. In polymerisation, another chemical is added, acting as a catalyst and causing the monomers to combine to form a resin. Polycondensation involves combining monomers so that they release a by-product such as water or alcohol. With either process, additives such as pigments can be mixed in to modify its properties.

Plastics fall into two categories: thermoplastics and thermosets. After being moulded and cooled, thermoplastics can be melted and remoulded because their polymer chains bond weakly, and are easily broken by heating. Heat a thermoset, on the other hand, and its polymer chains bond in a way that cannot be undone – it is permanently 'set' into shape and can't be melted again.

PVC unwrapped

The polyvinyl chloride polymer at an atomic level

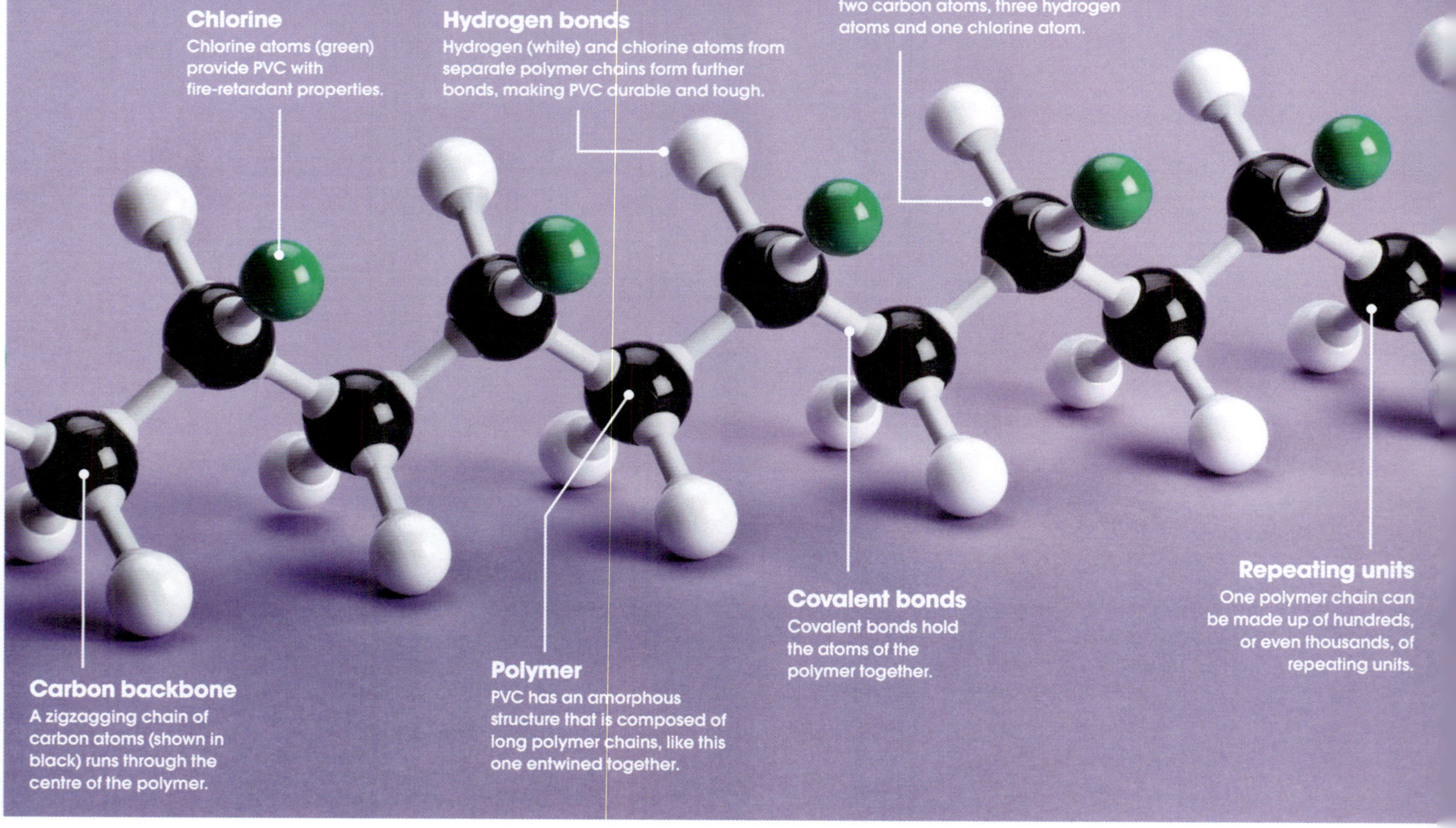

Producing and recycling plastics

To make a product, manufacturers melt plastic pellets and use one of several techniques to mould the plastic into its final shape. Plastic tubs, for example, are made by injection moulding, where heated pellets are injected into a mould cavity at high pressure. Films and piping, in contrast, are created by forcing the molten plastic through a small opening – a method that is known as extrusion.

Recycling plastics is, in theory, quite straightforward: grind the material down into flakes, wash away impurities, melt it back down into pellets and you're ready to start again. In practice it's a little more complicated. While thermoplastics such as PET can be melted down and reused, thermosets do not melt and cannot be recycled. Different types of plastic need to be sorted and processed separately, which is a labour-intensive process. Other plastics, such as polystyrene, are simply not cost effective to recycle.

n the future

ne plastics and polymers
nder development

Vith oil and gas reserves declining, the ressure is on to develop viable ioplastics, made from plants such as orn or cane sugar. More research is eeded, however, to bring down their ost and resolve the environmental hallenges they pose when it comes to iodegradability and recycling.

Self-healing polymers, meanwhile, ould spell the end of scratches on our phone's screen. Different methods xist, one being to add microcapsules f a 'healing agent' to a polymer ructure. When the plastic is amaged, the healing agent oozes ut, catalysing a reaction which bonds back together.

Elsewhere, 3D printing continues to pen up many new possibilities for what ou can do with plastics. Instead of ink, 3D printer head is filled with a polymer r sometimes a metal) which it melts nd prints in layers to create all manner f three-dimensional objects, from coat ooks to human bones for transplant.

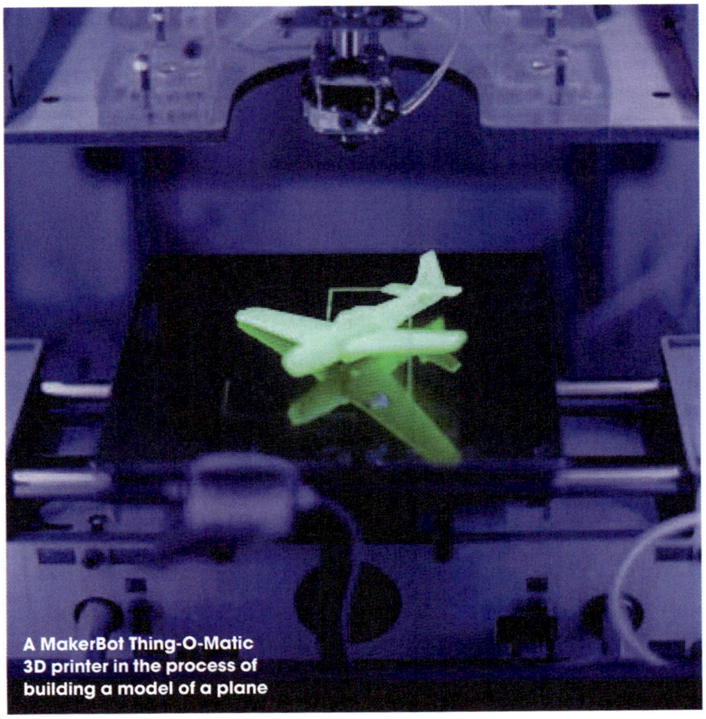

A MakerBot Thing-O-Matic 3D printer in the process of building a model of a plane

Plastics for all purposes

Polyvinyl chloride (PVC)

PVC's durability and resistance to chemicals, grease and oil make it ideal for plumbing. It can be manufactured into rigid or flexible forms.

Polystyrene

Manufactured into a foam, polystyrene is an excellent heat insulator. Polystyrene chips or trays also absorb impacts to protect delicate goods.

Low-density polyethylene (LDPE)

Soft and flexible, LDPE is strong even when processed into thin sheets – so it's perfect for shopping bags. LDPE's molecular structure also gives it some stretch, making it suitable for food wraps.

Polyethylene terephthalate

A mainstay of the food and drink industry, PET is tough and impermeable to water and most gases. PET containers don't release chemicals into the food or fluid they contain.

Polypropylene (PP)

Polypropylene is strong yet flexible. Able to withstand higher temperatures than most plastics, it can be washed in the dishwasher or even heated in the microwave without warping.

13 Chemistry life hacks

A little science knowledge goes a long way when it comes to solving these everyday problems

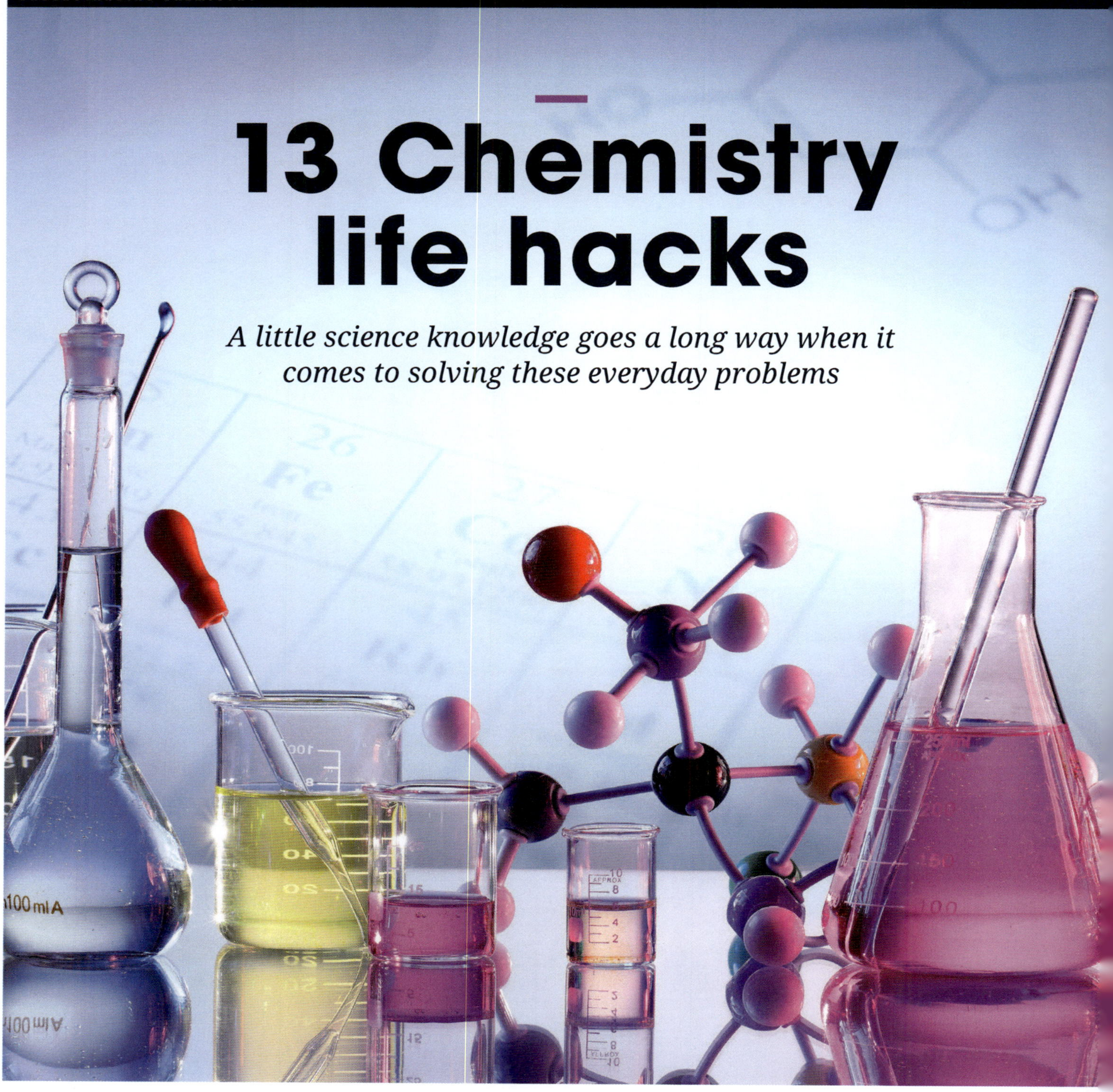

A common complaint in science classrooms is that nothing you learn in your school chemistry class will actually be useful in everyday life, but this doesn't have to be the case. Put simply, chemistry explains what matter is made of, the properties it has, and how different molecules interact in the world around us. An understanding of the chemistry behind common home conundrums can be used to make your food taste better and save you a ton on expensive cleaning products.

We've got 13 chemistry life hacks that demonstrate this, including how getting to grips with freezing points can cool your drink in record time, how electrolysis can clean your silver jewellery, and how understanding acids and bases can de-stink your fridge. We've also got a trick for removing rust with an everyday acid, a way to use your knowledge of solubility to get a red wine stain out of your carpet, and a food chemistry hack for reviving stale cookies.

Fix bitter coffee

It's well known that salt makes food taste good, but in 1997, scientists found out why. In a study published in the journal *Nature*, they asked volunteers to drink a bitter urea drink with added sucrose for sweetness, or sodium acetate for saltiness. Sugar didn't improve the taste on its own, but salt did, and when the two were combined, the drink tasted even sweeter. It seems salt blocks bitterness and boosts sugariness.

Coffee contains many different bitter-tasting molecules, which contribute to its very distinctive flavour, but it takes a careful balance of temperature, brew time and ground size to get the perfect cup. It's very easy to get this delicate balance wrong, often with not so tasty results, but salt's ability to suppress bitter tastes and boost sweetness could be just the thing you need to improve your brew.

"Salt blocks bitterness and boosts sugariness"

Coffee chemistry

The smell and taste of coffee are the result of a complex mix of molecules

Bitter molecules

Coffee beans naturally contain chlorogenic acids, which break down into bitter quinolacetones, phenyl indanes and melanoidins when the beans are roasted. These are responsible for the flavour.

Brew time

The longer the coffee grounds are in contact with water, the more the bitter compounds enter the drink and the stronger it tastes.

Water temperature

The temperature of the water affects how well the bitter molecules are extracted from the coffee. Ideally, it should be between 90 and 96 degrees Celsius.

Grind size

The finer the grind, the more surface area is in contact with the water and the faster the bitter molecules will dissolve.

Salt

Adding salt counteracts the bitterness and increases sweetness. Minerals in hard water affect the taste, too.

Remove rust with cola

Phosphoric acid is the sour ingredient that balances the sugary taste of the world's favourite fizzy drink, but it's got another use – it's an industrial-strength rust remover capable of transforming reddish iron (III) oxide into yellow-toned ferric phosphate. To repair a rusty object, simply cover it in cola and then let the acid get to work. Not only will it help to remove the rust, the layer of iron phosphate will actually provide some rust proofing, protecting against future corrosion.

Iron reacts with oxygen in the presence of moisture, creating red-toned iron oxide

Check if eggs are fresh

Fresh eggs are full to the brim with yolk and white, but the shell is porous to allow the developing chick to breathe. Over time air leaks through the protective coating and a bubble starts to form. For a sure-fire way to tell if your eggs are fresh, simply put them in a bowl of water and see if they sink or swim.

The freshest eggs are best for frying or poaching because the yolk is round and the white is thick. These should sink to the bottom of the bowl and lie down horizontally on their side. Slightly older eggs are better for hard boiling because the white is thinner and they are easier to peel. These eggs will tip up on their edge, sitting upright in the water but not quite floating. The oldest eggs will rise to the top of the water. These are best thrown away as they are not really fit to eat.

Fresh or not?

How to quickly identify a good egg without having to crack the shell

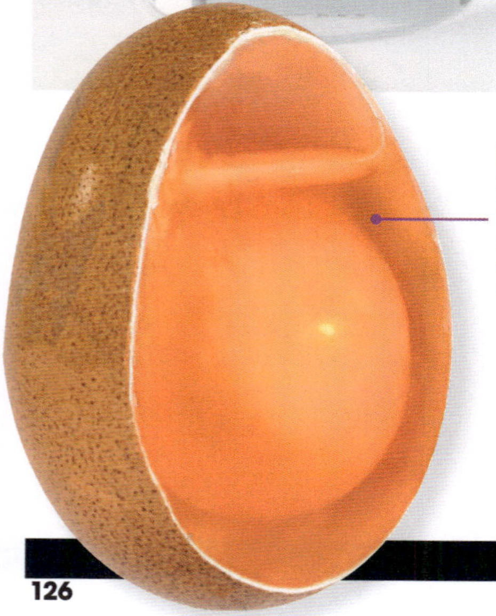

"Eggshells are porous to allow developing chicks to breathe"

Cool your drink with salt

There's a layer of liquid water on the surface of ice, and when you add salt it dissolves into this layer. This dissolved salt slows down the rate at which water molecules can attach to the ice beneath but doesn't affect the rate at which molecules detach from the solid. The overall effect is more molecules detaching and becoming liquid, allowing the ice to melt without raising the temperature.

If you put a can into an ice bucket, only parts of the surface touch the chilly solid, but add salt to the mix and the chilly liquid will completely surround your can, cooling it down much faster.

Gritters coat the road in salt to lower the freezing point of ice and snow

Keep your veg green

When you first plunge veggies into boiling water, air bubbles expand, making their green colour appear more intense, but over time the vibrant shade turns yellow. This is because a magnesium ion in the green pigment, chlorophyll, is swapped for hydrogen in the presence of acids released during cooking. To prevent this from happening, minimise cooking time, use lots of water to dilute the acid, and don't add vinegar or lemon until you're ready to eat.

Plant cells are stuffed with chloroplasts, which contain the green pigment chlorophyll

Ripen fruit with fruit

ipe fruits like tomatoes and bananas ive off ethylene, or ethene. This natural hemical is also known as 'fruit-ripening as' and has been used for centuries to rtificially speed up fruit ripening. The gyptians made holes in figs to let the gas ut, people in ancient China burnt ncense to release ethylene, and modern ransport vehicles pump the gas over fruit n their way to the supermarket to ensure 's ready to eat on arrival. To use this trick t home, place an ethylene-producing ruit, such as a banana, fig, mango, ectarine or plum, next to an unripe fruit nd it'll be ready to eat in no time.

favourite of the ancient Egyptians, gs encourage other fruits to ripen

Thread needles with nail varnish

Poking a soft, frayed thread through the eye of a needle can be a challenge, but a bottle of nail varnish makes the job much easier. Nail varnish contains nitrocellulose suspended in a fast-evaporating solvent – when you apply a dab to the end of your thread, it rapidly forms a smooth film over the strands.

Bicarb hacks

Humble bicarbonate of soda is all you need to solve these common household problems

Refresh tarnished silver

Silver can lose its lustre over time, becoming coated in a layer of black silver sulphide, but there's a chemistry trick to gently restore its shine. The first thing to do is line a bowl with aluminium foil and then add a spoon of bicarbonate of soda, a pinch of salt and some hot water. The bicarbonate of soda will react with the foil, stripping away the layer of aluminium oxide on the surface, while the salt allows electrons to move between the foil and the silver, creating a small electrical current. The silver gains electrons and the aluminium loses them and, in the process, the sulphur transfers from your cutlery to the foil.

The black tarnish on old silverware is silver sulphide

Make bread without yeast

The bubbles that make bread dough light and airy are most often made by yeast, a single-celled organism that consumes sugar and produces carbon dioxide, but you can still make a decent loaf without it. Soda bread is a traditional quick bread with four key ingredients: flour, bicarbonate of soda, salt, and an acid like buttermilk, milk, yoghurt or cream of tartar. The soda reacts with the acid to make carbon dioxide, allowing the bread to rise. But be careful with the proportions; the bread will collapse if the bubbles become too big, and too much bicarbonate of soda can make the mix taste soapy.

Bicarbonate of soda can replace yeast to make dense, rustic soda bread

Banish fridge odours

Bicarbonate of soda has the chemical formula $NaHCO_3$, and it's got some special properties. It is amphoteric, which means it can react with both acids and bases, and this makes it a great DIY fridge deodouriser. Lots of bad food smells are caused by acids and alkalines produced as food starts to go off; sour milk contains lactic acid, bad meat contains rancid fatty acids, and rotten fish contains alkaline trimethylamine oxide. When bicarbonate of soda reacts with these, it forms a sodium salt, water and carbon dioxide, neutralising the odour. An open box inside the fridge door should help to keep bad smells at bay.

Bicarbonate of soda can neutralise bad smells by reacting with acids and alkalis

Remove wine stains with vodka

he distinctive colour of red wine is created y pigments called anthocyanins and yranoanthocyanins, which are formed when nthocyanins interact with molecules made y alcohol-producing yeasts. Their chemical tructure makes them a little bit hydrophobic 'water hating') and a little bit hydrophilic water loving'), meaning that they dissolve in oth organic solvents, like alcohol, and aqueous olvents, like water.

White wine is often used to remove red wine tains because it contains a mix of alcohol nd water, helping to capture the pigment nolecules and pull them out of the fabric, but better option to quickly remove a stain is to use something stronger. Clear spirits like vodka, white rum or gin contain a higher percentage of alcohol, dissolving the pigment molecules even more effectively. Just keep dabbing at the stain and adding more alcohol until the colour begins to fade away.

Carpet stain chemistry

ow to dissolve red wine pigments before they stain your fabrics

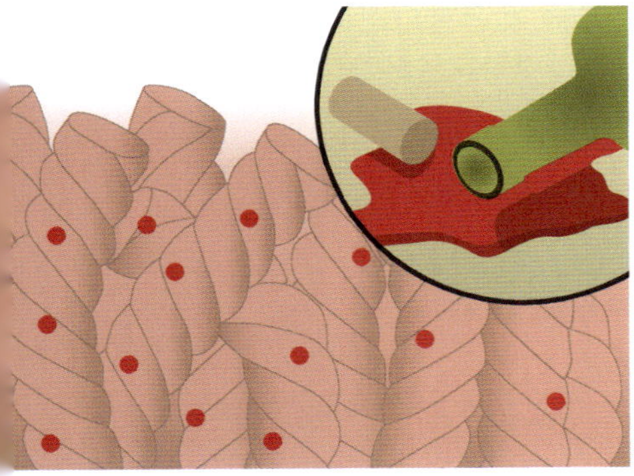

Stain

ne anthocyanin pigment molecules in red wine get uck between carpet fibres, staining the fabric pink.

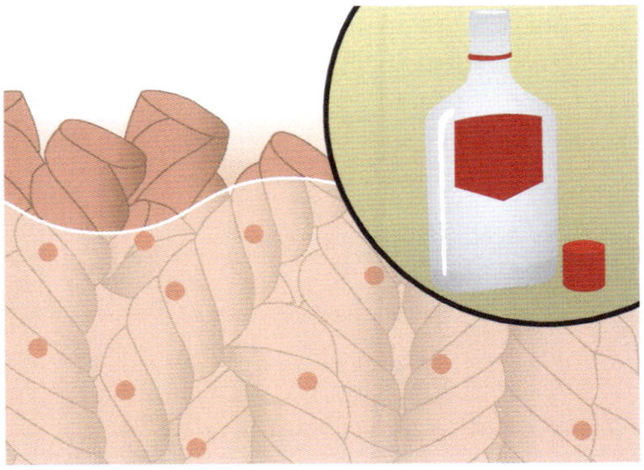

2 Add alcohol

The mix of water and alcohol in a high-proof spirit like vodka dissolves the pigment molecules.

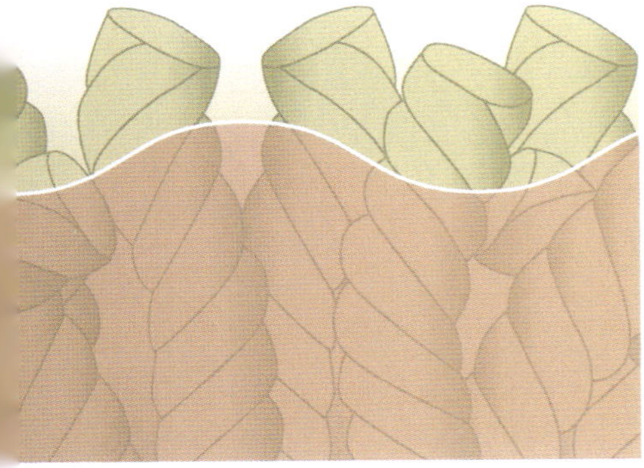

Dab

ne dissolved pigment can be removed from the arpet by dabbing at the stain with a cloth.

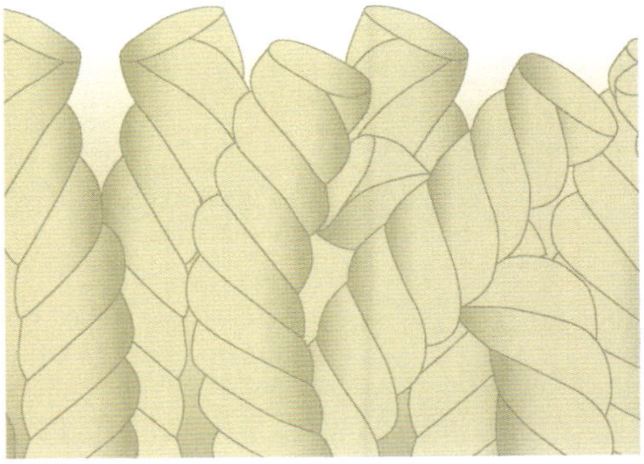

4 Rinse

Rinse any alcohol residue away with water and your carpet will be as good as new.

Descale your kettle with vinegar

Hard water contains dissolved calcium hydrogencarbonate, which breaks down into insoluble calcium carbonate, also known as limescale, when it's heated. To get rid of it, mix one part vinegar or lemon juice with three parts water and boil the kettle, then let the hot mixture sit overnight. The acid will react with the limescale, forming soluble calcium salts that can be tipped away in the morning.

A scanning electron microscope image showing crystals of limescale inside a kettle

Soften cookies with bread

Bread, cake and cookies contain moisture, which transforms starch from crystals into a disorganised gel, but as moisture moves away from the starch, they go hard and stale. All you need to do to refresh baked goods is restore that moisture. Simply put your hard cookie in a bag with a piece of fresh bread and it should soften up in no time.

The starch inside cookies turns to orderly crystals as they dry out

HOW IT WORKS
The Collection
Future Genius
Human Body
Issue 3
NEW
Anatomy Cells Nutrition Your genes + More!
HOW YOUR BODY WORKS
INCREDIBLE HISTORY
The SPACE RACE
ROBOTS
SPACE
FEED YOUR MIND WITH OUR BOOKAZINES
Explore the secrets of the universe, from the days of the dinosaurs to the miracles of modern science!
AMAZING SCIENCE
Life's Little Mysteries
BLACK HOLES
WHAT MAKES A BLACK HOLE SUPERMASSIVE
THE Science COLLECTION
Future Genius
Ancient Egypt
UNDERSTANDING HABITS
HOW IT WORKS BOOK OF THE HUMAN BODY
Follow us on Instagram @futurebookazines
FUTURE
www.magazinesdirect.com
Magazines, back issues & bookazines.

UNDERSTANDING Chemistry

Future PLC Quay House, The Ambury, Bath, BA1 1UA

Editorial
Editor **Alice Pattillo**
Senior Art Editor **Andy Downes**
Head of Art & Design **Greg Whitaker**
Editorial Director **Jon White**
Managing Director **Grainne McKenna**

How It Works Editorial
Editor **Ben Biggs**
Group Editor-in-Chief **Tim Williamson**
Senior Art Editor **Duncan Crook**
Production Editor **Nikole Robinson**
Senior Staff Writer **Scott Dutfield**
Staff Writer **Ailsa Harvey**

Cover images
Getty Images

Photography
All copyrights and trademarks are recognised and respected

Advertising
Media packs are available on request
Commercial Director **Clare Dove**

International
Head of Print Licensing **Rachel Shaw**
licensing@futurenet.com
www.futurecontenthub.com

Circulation
Head of Newstrade **Tim Mathers**

Production
Head of Production **Mark Constance**
Production Project Manager **Matthew Eglinton**
Advertising Production Manager **Joanne Crosby**
Digital Editions Controller **Jason Hudson**
Production Managers **Keely Miller, Nola Cokely, Vivienne Calvert, Fran Twentyman**

Printed in the UK

Distributed by Marketforce – www.marketforce.co.uk
For enquiries, please email: mfcommunications@futurenet.com

Understaning Chemistry First Edition (HIB5754)

FUTURE
Connectors. Creators. Experience Makers.

Future plc is a public company quoted on the London Stock Exchange (symbol: FUTR)
www.futureplc.com

Chief Executive Officer **Jon Steinberg**
Non-Executive Chairman **Richard Huntingford**
Chief Financial and Strategy Officer **Penny Ladkin-Brand**

Tel +44 (0)1225 442 244

Widely Recycled

ipso. For press freedom with responsibility

Printed in Dunstable, United Kingdom

91222488R00076